Six Sigma, Basic Steps & Implementation

By

Fred Soleimannejed

First published by AuthorHouse 04/29/04

ISBN: 1-4184-5744-2 (e-book)
ISBN: 1-4184-4801-X (Paperback)

This book is printed on acid free paper.

Abstract

Six Sigma

For the last several decades the quality improvement concept has been part of every executive's agenda to work on. Hence, the quality gurus have introduced the new concepts to response to the customer's needs and requirements. Among the new ideas and programs, Six Sigma has become synonymous with defect free low cost performance in the hands of blue-chip companies such as GE and Motorola. Its power comes from the relentless focus on the Voice of the Customer (VOC) and continuously defining and improving on the processes that serve them until delivery is flawless. It requires a new set of measurement techniques in order to comply with the progress of industry.

As the industry progresses the control of process becomes more sophisticated and critical for the manufacturers as well as the customers. "Defect free product" is a very common term in this industry. In order to achieve that level, we need to design a robust process which could respond to the demanding requirements. Motorola, Inc. was one of the first companies to introduce the Six Sigma concept as a tool for measuring the manufacturing processes.

For those who are familiar with statistics, sigma is the measure of deviation from a mean value. In business, variation in a process (whether that process is the machining of a diameter or the collection of overdue receivables), almost always results in defects, rework and scrap. If one can reduce the variation in a process, one reduces the defects of that process. But Six Sigma, as implemented by Motorola, GE and others, is much more than an abstract statistical concept. Stated simply, Six Sigma is a methodology to eliminate defects and to eliminate the opportunity for defects. A more complete understanding reveals that Six Sigma includes:

1.   Business strategy and vision;

2. Standard set of metrics to measure improvement;

3. Benchmark to compare performance;

4. Philosophy of how to pursue improvement;

5. Methodology to measure, analyze, improve and control the business processes;

6. Comprehensive set of tools, including both "hard" statistical tools;

7. The setting of aggressive goals.

This book identified the basic steps for defining and implementing Six Sigma.

Acknowledgment

Completion of this book would not have been accomplished without the contributions of my wife Anahita. I would like to recognize her efforts and offer my sincere appreciation for all she has done. I would also like to thank Chris Mrena for editing and her advise.

To: Sena

Table of Contents

1 CHAPTER ONE

1.1 *Historical Overview*

For the last several decades we have experienced an enormous explosion of technology. Because of this explosion, we see products that operate faster, have more features, occupy less space, and in some instances, even cost less than their predecessors with higher quality do. At the same time, quality improvement has been one of the main topics of industry. Quality gurus have developed different philosophies such as Total Quality Management (TQM) and Total Quality Control (TQC) to achieve and maintain quality with the explosion of technology.

There has been lots of books published and have explained the theory of Six Sigma and implementation. In this book the approach is quite simple and deals with the basic elements of the program. It does not claim hat the Six Sigma is the ultimate and the only way to resolve organizational issues and improve the processes and satisfy the customers. Six Sigma is one of the tools which starts with the customer's perspective to the suppliers quality and service and from this point on the road map will be developed and finally leads to the organizations' profitability and customer satisfaction. It does not mean that the Six Sigma approaches are the most appropriate one for any and all the organizations. One of the main reasons that the company decides to implement the program is to enhance and improve the bottom line profit margin.

The Six Sigma philosophies were developed during the 1980s when more accurate measurement techniques were required to quantify quality improvement. Also, the new concept of process control was adopted in the mid-80s; that was a program supporting to achieve the highest level of quality through the process optimization and Six Sigma.

1

Sigma is a Greek word used to describe variability. The classical measurement unit for the Six Sigma program is "defect". The Six Sigma quality level is designed to measure the processes based on the "Defect Per Million Opportunity" DPMO. The higher the sigma is the higher the quality will be. From the statistical perspective it's an advance statistical analysis approach to data. It also disciplines the organization to practice numbers and facts in order to make any decision based on the subjective ideas. Unfortunately there has been a misunderstanding of the statistical approach as the only dimension of Six Sigma. The most powerful aspect of the program is creating and developing strategy for the business and overall for the organization. The combination of using statistical approach and business strategy has made Six Sigma so powerful, which can resolve almost any organization problems. It enables the management to make the critical decisions based on facts and figures. Kim Fudge, manager of operations at Motorola notes in Gill's (1990) article, "operating at Three Sigma smacks of surrender" that the data which has been collected from the three sigma process is not sufficient to make a critical decision.

According to the American Society for Quality Control (1991), AT&T lab conducted an analytical experiment on the quality and productivity of their organizations. They realized that improvement of process capability by adopting the Six Sigma philosophies is the key to total customer satisfaction. They believe that quality and reliability are the most important factors in the manufacturing and safety of products. Achieving the highest level of safety and reliability requires a robust quality system, which is quantifiable and can effectively prevent any opportunity for error.

Motorola University Press (1993) conducted an experiment on the realistic specifications and its relationship with Six Sigma capabilities. They noted that a great deal of work has to be done to come up with realistic specifications. Finding the best compromise between the customer's need and the manufacturability of a process is often a real problem. Since the process variation width is usually greater than gap between the lower and upper specification limits, extensive outgoing

2

inspection was needed to guarantee a defect rate of less than 1%. With the adoption of basic control charts and Pareto analysis, progress was made during the eighties. Three Sigma results in a product nonconformance level of 2700 PPM.

During the 1980s, McDonald Douglas expressed their concerns about improving quality by creating a comprehensive statistical program. McDonald Douglas, through the Six Sigma researches, envisioned a collection of tools that would accelerate the achievement of quality within the company and facilitate the transfer of this knowledge to the technical and managerial communities. Process characterization and optimization and the others in the Six Sigma researches were designed to provide statistical solution to their quality improvement challenges.

Having 2700 defects per million or operating the process at the Three Sigma Level may be acceptable and cost effective for some industries, but certainly it is not acceptable for industries that are involved with the lives and safety of people. Twenty-seven hundred crashes after one million flights will stop anybody from traveling by airplane, or producing twenty-seven hundred defective doses of medication out of a million can cause twenty-seven hundred people to die. Measuring the processes by part per million (PPM) is a precise measurement, which takes into account almost all of the probabilities for error.

For the reasons mentioned above, the importance of a product requires very sensitive methods and measurements that respond to discrepant materials or products. It means new tools are required to measure the components of a medical device and control the behavior of its manufacturing process.

Allderdge (1993) describes Motorola's definition of quality. He explains that "in 1985 we attempted to define quality, but found ourselves too often pursuing quality that was not needed. A gold-plated restroom is not cost effective, nor is completely germ-free floors in an industrial setting, nor can we afford to remove every blade of grass that is not green. To use a recent phrase by Heinz Goodman (1990) a consultant on sales and marketing quality, The only quality is that which a customer is willing to pay for. No Company will survive long by spending for quality the customer is indifferent about. In the beginning Six Sigma seemed to be very costly and expensive but after a while we realized that we have saved lots of money".

Michael R. Tobin (1993) explains 'that controlling process through a statistical approach has been an established practice for a long time. The output parameters do not interest most industries. The interest has shifted to input parameters that assist in preventing failures from occurring. Thus, meeting specification is no longer the measure of how well processes are doing. The new wave of thinking is to control the process at the center of the specifications to reduce the probability of a nonconforming part being manufactured even if the process has shifted slightly from the normal behavior.

Another perspective of this study is to understand and evaluate the variation. The goal is to reduce and minimize variation. Thus, the source of variation needs to be Identified and controlled.

Motorola was the first company to introduce the Six Sigma new philosophies and to begin to measure their manufacturing operation by using the Six Sigma concept. According to USA Today (1989), Motorola engineered one of the most dramatic comebacks of this decade. In the final analysis, such performance was in great part, attributed to a corporate-wide caused for quality. President Reagan at the White House formally acknowledged this accomplishment in 1988, when he presented the first Malcolm Baldrige National Quality Award to Motorola Inc.

1.2 Definitions of Quality

J.M. Juran (1992) defines quality as a Frequency of Deficiency divided by Opportunity of Deficiencies. He also defines opportunities for error from the design to manufacturability of the product. O. R. Wade (1967) defines, total opportunity as design plus materials plus process. He also indicates that in order to implement the Six Sigma concept we need to consider all the opportunities, which are involved with these elements. Patrick O'Conner (1991) describes the product design as the first step towards achieving the Six Sigma program. Mikel J. Harry and J. Ronald Lawson (1992) emphasize on the source of variation for any given process. They believe that many variables within any manufacturing cause system can be classified into three primary source of causation. The sources are (a) inadequate design margin, (b) insufficient process control, and (c) unstable material and components. The Six Sigma takes all these variable into consideration.

The Research Institute of Motorola (1992) describes, the precision of the Six Sigma program and the improvements that Motorola has made by implementing this program. Motorola Semiconductor Products Sector (1992) specifies the criticality of the different products and how to take this critical factor into consideration.

M. Wapole (1989) stated the probability and the distribution function of each process and refers to the Design of Experiment as a key factor to identify the source of variations. R. Hogg and A. V. Gary (1989) introduced the generating function as a key to understanding different distributions such as the Poisson distribution. They also describe the Central Limit Theorem and explain its correlation with any distribution function. Donald Campbell and Julian Stanley (1963) stated, "by experiment we refer to that portion of research in which variables are manipulated and their effects upon other variables observed".

Koons (1993) explains the application consideration of the Six Sigma program. He says: "As with any other statistically based measure of quality, there are certain practical aspects to consider if the technique is to be correctly applied and the results are not to be misleading. The calculated index of the Six Sigma depends on estimates of the average and standard divination. Sample selection has an effect on these estimates".

According to Harry and Lawson (1994), the yield method is predicted upon knowledge of first time yield. Again, notice that first times yield is designed to measure the overall performance of the process. The defect per million opportunity is an indicator to show the details of the process.

Lawson (1992) suggested that the Process capability indices can be calculated for a wide variety of engineering, manufacturing, and administrative situations. Samples of five items were drawn from the process's output stream on five different occasions. The tabulated data are coded values; actual values, of course, would depend on the specific performance characteristic. The coded values can conveniently be thought of as deviations of the performance characteristic from its target value. In practice, a larger sample should be used to ensure meaningful results. In practice, a larger sample will be needed to ensure meaningful results. As previously indicated, we must take into account naturally occurring sources of manufacturing variation in order to define the nature of a $\pm 6\sigma$ characteristic; however, when the mean variation assumes the form of a sustained static shift we must handle the variation in a slightly different manner.

As with any other statistically based measure of quality, there are certain practical aspects to consider if the technique is to be correctly applied and the results are not to be misleading. some of the practical aspects of the Six Sigma capability metrics are given in this section. The calculated index depends on estimates of the average and standard deviation. Sample selection has an effect on this estimates. If measurements are taken over too short a period, that standard deviation may be understated.

This method involves gathering empirical data over many periods of production. It should be recognized as the method of preference in terms of analytical precision; however, it is perhaps the most costly and time consuming. As one may readily surmise, this method relies heavily on the use of statistical process control (SPC) charts for continuous data. The principal aim is to establish that a reliable estimate of σ has been sufficiently estimated, σ may be computed, assuming σ has already been estimated. At this point, k may be calculated. The reader should be aware that certain SPC charts can provide an estimate of σ in a single be stimated using the same chart.

Koons (1992) concluded, It must also be recognized that a detailed presentation of this particular method is far beyond the scope and intent of this research. There are simply too many charting alternatives to consider. Because some of the alternatives are fairly sophisticated in terms of they're underlying statistical theory, the novice researcher should seek the advice of an experienced practitioner prior to chart selection and application. Should additional information on SPC charts are desired, the reader is directed to the bibliography.

A study conducted by Motorola (1992) examined the effectiveness of the Six Sigma program for different industries. They concluded that, at the beginning the cost of implementation of the program would be expensive but, as much as the program progresses this cost will reduce and the outgoing result will be more favorable.

Gill (1990), author of "Staking Six Sigma," notes examples of what having %99 quality means. It is stated that:

1. The US Postal Service would lose 17,000 pieces of mail every day.

2. More than 30,000 newborn babies would be dropped accidentally in hospital every day.

3. 5,000 incorrect surgical operation per week is performed.

4. Two short or long landing is made at most major airport each day.

5. 10,000 defective Printed Circuit Boards are produced per each million.

As pointed out earlier, the 5-year plan "Six Sigma by 1992" was launched in 1987 to reach quality standards approaching zero defects in everything Motorola does, including Manufacturing, Sales, and Communications. Six Sigma applies equally to any process at Motorola, such as processing orders or preparing financial statements.

Motorola's Corporate Office (1987) has documented a learning guide on Six Sigma entitled what is Six Sigma? The guide is an educational tool given to employees which addresses basic concept about the normal distribution of processes. It indicated that approximately 2,700 parts per million will fall outside the normal ± 3sigma variation. At first glance, it does not appear to be disconcerting, as there are still 997,300 parts, which are within the ±3 sigma limits. However, as Motorola's what is Six Sigma? points out, when a product is built containing 1,200 parts, it can be expected that the compounding of the defective parts in, on average, 3.24 defects per end product (1,200*0.0027). This represents a cumulative yield for the entire manufacturing process of less than 4%. This percentage correlates to fewer than 4 units out of every 100 units manufactured would ever go through the entire manufacturing cycle without a defect.

Harry (1992) explains that the Six Sigma concept is a relatively new way to measure how good a product is. When a product is Six Sigma it tells us that product quality is excellent. It says the probability of producing a defect is extremely low. It is also a target that Motorola Inc. has established for the quality of its products. To better understand how Six Sigma tells us that a product is excellent, let the term "Sigma" acutely means in term of quality. Essentially, a sigma is a statistical measuring device that tells us how good our product are .Using this device we can directly measure the quality confidence we would have in a given product or process and then compare it to another like product or process. To make things simple, will substitute the symbol "σ" forth word "sigma." To apply this concept, we must first determine how many opportunities for error exist as related to a particular product. Next, we must count the actual number of defects associated whit that product during manufacture. For example as it will be explained for the Printed Circuit Assembly Industry (PCBA) the number of solder joints determines the actual opportunities for error. With this information we are now able to determine how many defects there are per million opportunities for a defect. For example, if there are 1,000,000 opportunities for a defect to occur within each of our five board and we observe five defects (one defect pre board) then there would be one defect per million opportunities or, expressed as a fraction, 0,000001 nonconformities per million opportunities (NPMO). Note that this also maybe expressed as parts per million (PPM). The abbreviations NPMO and PPM are used interchangeably.

Now, the last thing that we must do is statistically convert the NPMO in to sigma units. When the number of sigma units' σs is small, say two, product quality is not very good. The number of defects per million opportunities for a defect would be intolerable. When the number of σs is large, say six, quality would be excellent. The number of defects per million opportunities would be extremely small. In this sense, the quality σ's are like water graduations on a glass beaker -like chemists use to measure the volume of a fluid. So think of quality like water in a beaker -the more quality we have in the more fluid there is in the beaker.

In general, when we say that a product is 6σ, what we are really saying is that any given product exhibits no more than 3.4 NPMO at the part and process step levels. This number (3.4) takes typical sources of variation into account. That is to say the 6σ quality concept recognizes that a small amount of variation will be present as a result of slight fluctuations in environmental conditions; differences between operators, parts, materials; and so forth. If the fluctuations can be sufficiently controlled such that a product or process characteristic stays centered on its ideal condition, there would be only 0.002 nonconformities defects per million opportunities. However, as we are aware, nothing can be perfectly controlled to an ideal condition - some shifting and spread will be present over the long haul. When such variation is taking into account, there would only be 3.4 NPMO this is an exceptionally high level of product quality-the competitive edge.

Harry (1992) describes that since Motorola Inc. first introduced the 6σ program, people have been asking why the various chart, graphs, diagrams, paper, and presentation on the topic indicate that 6σ is equivalent to a defect rate of 3.4

parts per million (PPM). The broad answer to this question would be that the definition of 6σ, as established by the corporation assumes a 1.5σ shift in the population average. More will be said a bout this assumption later on in the discussion. When such change is taken into consideration, the end result is a defect rate of 3.4 PPM, as compared to 0.002 PPM. Thus, 0.002 PPM, represents the steady -state level and 3.4 PPM denotes the dynamic "real world" state of affairs, with regard to 6σ quality performance. To fully appreciate the intent and purpose of the 6σ program, we must get a little more technical with our explanation. First of all, we must recognize that virtually nothing in this world is static. We live and work in a very dynamic environment where almost every thing is changing. For instance, within our production areas, such variables as room temperature and humidity are constantly changing. Even the tooling we use incurs wear over time. The smallest change in such things as raw materials and parts, causes product performance, quality, and reliability to vary- sometimes for the better. In turn, such variation can significantly impact things like cycle time and inventory. As a consequence of variation, things become less predictable and thus, much harder to manage. The repercussions of variation are felt throughout an organization. Ultimately, the customer feels the effects of variation. So what is to be done. Perhaps the answer can best be given by an example.

1.3 What is Six Sigma?

Six Sigma is a disciplined methodology for improving organizations' processes, based on rigorous data gathering and analysis. The approach focuses on helping organizations produce products and services better, faster and cheaper by improving the capability of processes to meet customer requirements. Six Sigma identifies and eliminates costs, which add no value to customers. Unlike simple cost-cutting programs, however, Six Sigma delivers cost cuts whilst retaining or improving value to the customer.

The term Six Sigma is based on a statistical rationale. Six Sigma performance is the goal and equates to 3.4 defects per million process, product or service opportunities. The focus is on reducing variability to achieve the goal.

1.4 Six Sigma and Process tolerance

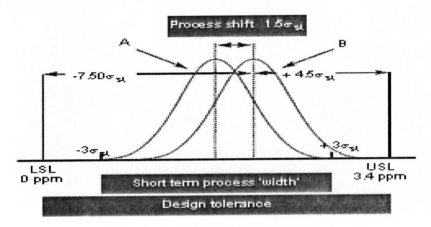

*"Figure1"*six sigma process tolerance

Figure 1 illustrates the Six Sigma scenario. The output from most processes follows a normal distribution as shown by curve A. In the short term, a process naturally centered at target and it will have a natural spread of plus or minus three-sigma. In the Six Sigma cases, this process variation is only half the width of the design tolerances for the process, i.e. the difference between the upper specification limit (USL) and lower specification limit (LSL). 99.9973 percent of the process output is contained by this natural spread.

Controlling processes in the longer term so that they remain on target can prove difficult in many practical situations. A typical process is likely to deviate from its natural centering position by up to one and half standard deviations. This is illustrated by a process shift to position B in Figure 1. Applying this principle, you can calculate the longer-term capability of the process based on the initial short-term capability of the process based on the initial specifications. The statistical basis

11

of Six Sigma, in particular the 1.5 sigma shifts, is not without its critics but this is outside our scope. Suffice to say the issue of process shifts and process capability is a practical problem in many industries. Six Sigma offers one solution.

1.5 *Six Sigma and cultural changes*

One of the dilemmas that the organization usually faces with is to make any changes and adopt a new program. Six Sigma has been known as a major change for the organization.

Dealing with change starts with understanding of it. The first stage in managing change should be as natural as possible. As we probably have experienced quality initiatives usually start with the executive management. They are the ones who define the strategy for the business and well as responsibility for profit and losses.

Knowing how to deploy versus delegate quality management activities is critical.

Deployment requires participation by executives to determine strategies, identify goals, select teams, review progress, etc. Delegating quality management assignments to others at lower levels demonstrate to employees that these efforts are low priorities. This is often interpreted as "the executive management doesn't support quality," "is going through the motions," and the result is that no one takes it seriously. Six Sigma is an extremely healthy and productive cultural change that takes time to complete. It is not free. It requires resources and training, but customer satisfaction, quality products and services produce a significant return on investment. Almost as important is the satisfaction all employees have from being on a winning team and their pride in being part of such a company.

1.6 Sigma capability

The management usually reacts to the bottom-line numbers. It is very important to determine and understand what your cost of (poor) quality is and what would enable a major change or help your line of work. Since 6-sigma basically focuses on process quality, it falls into the category of a process capability (Cp) technique. Traditionally, a process was considered capable if the natural spread, plus and minus three sigma (a yield of 99.73%), was less than the engineering tolerance. A later refinement considered the process location as well as its spread (Cpk) and tightened the minimum acceptable so that the process was at least 4-sigma from the nearest engineering requirement. Six Sigma requires that processes operate such that the nearest engineering requirement is at least plus or minus 6-sigma from the process mean. This requires considerable scientific and testing actions – often thousands of tests are run on multiple variables to get an understanding of what's going on.

Table 1 shows the competitive position of Six Sigma in regards to PPM and the Sigma level.

Sigma	Parts per million out of specification	% out of specification	Comparative position
6	3.4	0.00034	World class
5	233	0.0233	Industry best in class
4	6,210	0.621	Industry average
3	66,807	6.6807	Lagging industry standards
2	308,537	30.8537	Non competitive
1	690,000	69	Out of business!

2 CHAPTER TWO

2.1 Six Sigma Methodologies

The Six Sigma breakthrough strategies involve a 'define-measure-analyze-improve-control' (DMAIC) methodology, broadly based on the Shewhart plan-do-check-act cycle (frequently attributed to Deming). The strategy takes an organization's key business processes through five phases to deliver breakthroughs in performance.

Six Sigma starts with some basic and logical problem solving methodologies. DAMIC: Define, Analyze, Measure, Improve and control.

DAMIC is a process for continued improvement. It is systematic, scientific and fact based. This closed-loop process eliminates unproductive steps, often focuses on new measurements, and applies technology for improvement.

2. DMADV: Define, Measure, Analyze, Design and Verify are the basic tools that Six Sigma has used. The concept is a common sense approach to problem solving process.

It means in order to solve a problem we need to define it first. The experience shows that people from one organization usually see the problems differently. It takes hours and days and lot of brainstorming in order to bring everyone on the same page and finalize a problem statement that everyone aggresses with it.

2.2 Define Phase

Define phase involves defining the scope and goals of the improvement project in terms of customer requirements and the process that delivers these requirements. It also describes the problems and its relation to customer and internal organization.

It starts with evaluation of the voice of the customer (VOC). The VOC comes from verity of different sources. Customer's data and information, survey, dashboard and scorecard. Evaluation of VOC requires some basic steps, which need to be taken properly.

Identify the customer and which segment that they belong to. The segment could be financially, geographically, technology wise, etc. The critically of this step should be recognized. Compiling all different data from different segment will results to a conclusion/solution, which would not be appropriate for all the customers.

Compile customer data and evaluate it. Not all the data that is received from the customers could be analyzed and translated to a solution for problems. Noise must be removed from the data and the knowledge of subject matter should be considered in the evaluation process. Following are the basic and the critical steps through the define phase:

2.2.1 Translate the voice of customer (VOC) to the critical to quality (CTQ)

In order for any organization to act on any issue, it requires the organization completely understand the issue. VOC comes from verity of the sources and the involvement of the key personnel is a must to translate it to the organizations' language and processes. Organization then will take action and prioritizes the CTQs and develops the roadmaps.

Organization interests must be considered and recognized. The whole idea to implement the Six Sigma program is to make the organization profitable. The key element of the organization interest is that the employees understand the business strategies and goals. At what price are we willing to satisfy the customer?

2.2.2 Charter development, Problem Statement

At this point the team needs to develop a charter for the project. The charter could include the problem statement that describes "what is wrong" and goal statement that define the "improvement objectives". Quantitative goals based on the data will be preferable.

Following are some of the key questions that must be answered when the problem statement is developed:

- Will the project save cost?
- Does the project involve a clearly-defined process
- Who are the customers?
- What is a "defect" in the process?
- Is the process measurable?
- Is this issue important to the business?
- Does the project have organizational support?

2.2.2.1 *Formulate problem statement*

As it was mentioned, the essential key in the define phase is to explain the problem statement correctly. It has been experiences that half way through the execution of projects, appears that the team members do not have the same understanding of the problem statement. The problem statement must be clear and measurable. For example 10% of shipments are received late by customers, leading

16

to customer dissatisfaction and loss of business to competition or, reduce the defect rate by 20%.

2.2.3 Project champion and stakeholders

Another important key for implementation of a successful project is the support of executive management. There have been lots of cases that the projects have not been successful or implemented correctly because of lack of executive management support. It is highly recommended for a team to obtain the champion's support and agreement prior to execution of any project.

2.2.4 Developing process maps

In order to visualize the entire process for the project, the process map needs to be developed. It enables the organization simplify a complex process and identify the cycle time and non value add steps in the process. Constructing the process map requires the SIPOC analysis. SIPOC (supplier, input, process, output, and customer) helps the team to picture the impact of the project on the organization. To develop a process map basically three questions need to be answered.

What the team thinks the process is?

What actually the process is?

What the team would like the process to be?

Review, validation and approvals: Finally it comes to a point that all these information and processes must be reviewed with the champion and validates the process map against the charter and get the champion's approval prior to the execution of the program.

2.2.5 Critical Goal

Selecting the right goals is critical to identify the issues that are important to the customer. The goals are categorized in the model using a general-to-specific format. The customer-driven goals are organized into major categories, logically grouped by issues that are customer focused. Example groupings include customer requirements that are critical-to-quality, critical-to-delivery, critical to-cost and critical-to-process. Critical-to-process goals typically target internal activities that support the attainment of the customer critical goals.

Following are the necessary steps that need to be taken at the define phase. Each step should be documented and correlates with customer's needs and business strategies.

There should be also criteria developed by the Six Sigma implementation team to define and identify the boundaries with each step.

- Project selection criteria
- Clear Problem Statement and a Mission Statement
- Team Charter
- Select Team Members
- Guidelines for Team

In order to assure the success of each phase, a check list must be developed to identify the areas that need to be covered. At the beginning of each phase the team defines the elements and the tools (How) that can be used to measure that particular element. Then the team will identify the deliverables for each element. The deliverables must be met at the end of each phase. The checklist could be developed in any format as long as meets the criteria for the project and DMAIC phases.

2.2.6 Define Phase Checklist

Define	How? Tools	Deliverables
Identify Customer and Project CTQs	Surveys, Interviews, Focus Group, customer feedback, QFD	The area where the project is intended to focus
Develop Team Charter	Project Definition tree (Drill down from Big Y's), Change management tools, VOC to CTQ translation matrix	Documented agreement on the specific problem and goal you're setting out to solve
Business Case		Importance of the project
Problem Statement		What the problem is?
Goal Statement		What does the end state look like
Project Y Draft		How is the process measured
Scope	In Frame/Out of Frame	Define boundaries of what to work on
Team and Roles	Stakeholder analysis	
Financial Opportunity		Total opportunity, not committed benefits/ soft and hard returns on Six Sigma (ROSS)
High level process map	SIPOC, Process map	High level process map helps clarify the project scope gets people grounded on where the project is focused

2.3 *Measure Phase*

Measure phase involves measuring the current process performance -input, output and calculating the sigma capability for short and longer-term process capability. During the Measure phase, we refine the work done in the Define phase, as we gather additional information. We validate the customers' specifications for the outcome of the process. We define the defect that needs to be eliminated or reduced and perform a measurement system analysis to ensure our ability to measure the process outcome accurately. We explore the capability of the current process to meet the customers' requirements, determining the current performance (baseline) and the entitlement (potential) of the existing process. We document the current process using process maps. We start to define potential factors that drive the outcome of the process using cause and effect diagrams and begin to collect input and outcome data related to meeting the customers' requirements. As suppliers focused on improving our processes, we must realize that customers' requirements generally have three aspects. Customers focus on delivery (on-time or responsive), price (value), and quality (functionality or accuracy). As suppliers, we must focus on cycle time, cost, and defects to ensure we meet all aspects of the customers' requirements consistently. It is critical that we measure our processes the way the customers measure them; this is the only way we can understand whether or not we are meeting the customers' requirements. For example, if the customer is interested in on-time delivery, we can not measure on-time shipping just because it is easier, it does not reflect the Customers' requirement.

The first steps that help us to clearly measure the current performance are:

- Identify all the possible processes that need to be measured. The metrics and measurements must reflect the customer's requirements. It also requires a base line in order to make it meaningful.

- Prioritize and select the processes and metrics that reflect customer's requirements. At this step the manageability of the project is essential. If the scope of a project is not manageable, the measurement will be meaningless.

- Gather the required data and define unit of measurements, opportunities for error and set the performance standard.

- Project data collection plan is the next step that needs to be considered. It means the risk, accuracy, sources of data and measurement analysis should be evaluated and required actions need to be taken.

Tools that are typically used in the Measure phase are:

- Process capability analyses

- Process maps

- Cause and effect diagrams and matrices

- Failure mode and effect analyses (Will be explained in Chapter 3, 3.3)

- Measurement system analyses

- Graphical techniques

2.3.1 Process Capability

After analyzing data it is essential to understand and calculate the process capability. Process capability and process performance index are the key measurement for evaluation of the existing variability and goal setting

The total $\pm 3\sigma$ range of process variation is called Process Capability. Defects occur when process capability exceeds the design tolerance or total specifications width. Also, in order to quantify the relationship of design tolerance and process variation an index known as process capability (Cp) has been devised.

Cp = Process Capability. A simple and straightforward indicator of process capability.

Cpk = Process Capability Index. Adjustment of Cp for the effect of non-centered distribution.

Pp = Process Performance. A simple and straightforward indicator of process performance.

Ppk = Process Performance Index. Adjustment of Pp for the effect of non-centered distribution. Following is an example of Cp-indices. (Breyfogle III, F. W., 1999)

Figure 3 illustrates the process capability index comparison.

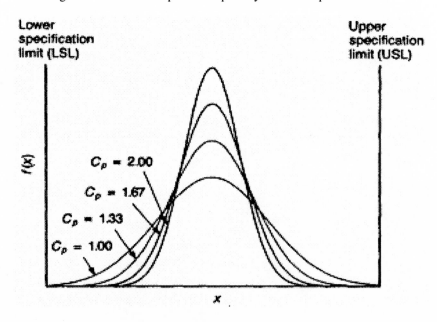

"Figure 3" Process capability comparison

2.3.1.1 *Interpreting Cp & Cpk*

"Cpk is an index (a simple number) which measures how close a process is running to its specification limits, relative to the natural variability of the process. The larger the index, the less likely it is that any item will be outside the specs. "If you hunt our shoot targets with bow, darts, or gun try this analogy. If your shots are falling in the same spot forming a good group this is a high Cp, and when the

22

sighting is adjusted so this tight group of shots is landing on the bull's eye, you now have a high Cpk."

Cpk measures how close you are to your target and how consistent you are to around your average performance. A person may be performing with minimum variation, but he can be away from his target towards one of the specification limit, which indicates lower Cpk, whereas Cp will be high. On the other hand, a person may be on average exactly at the target, but the variation in performance is high (but still lower than the tolerance band (i.e. specification interval). In such case also Cpk will be lower, but Cp will be high. Cpk will be higher only when meeting the target consistently with minimum variation.

We must have a Cpk of 1.33 (4 sigma) or higher to satisfy most customers.

Consider a car and a garage. The garage defines the specification limits; the car defines the output of the process. If the car is only a little bit smaller than the garage, you had better park it right in the middle of the garage (center of the specification) if you want to get all of the car in the garage. If the car is wider than the garage, it does not matter if you have it centered; it will not fit. If the car is a lot smaller than the garage (Six Sigma process), it doesn't matter if you park it exactly in the middle; it will fit and you have plenty of room on either side. If you have a process that is in control and with little variation, you should be able to park the car easily within the garage and thus meet customer requirements. Cpk tells you the relationship between the size of the car, the size of the garage and how far away from the middle of the garage you parked the car.

The value itself can be thought of as the amount the process (car) can widen before hitting the nearest spec limit (garage door edge).

Cpk=1/2 means you've crunched nearest the door edge

Cpk=1 means you're just touching the nearest edge

Cpk=2 means your width can grow 2 times before touching

Cpk=3 means your width can grow 3 times before touching

The general rule of thumb states that if a Cpk value of a process is less than 1.33, then the process is incapable of producing a repeatable part.

2.3.1.2 *Interpreting Pp & Ppk*

Process Performance Index basically tries to verify if the sample that you have generated from the process is capable to meet Customer CTQs (requirements). It differs from Process Capability in that Process Performance only applies to a specific batch of material. Samples from the batch may need to be quite large to be representative of the variation in the batch. Process Performance is only used when process control cannot be evaluated. An example of this is for a short pre-production run. Process Performance generally uses sample sigma in its calculation; Process capability uses the process sigma value determined from either the Moving Range, Range, or Sigma control charts.

2.3.1.3 *Differences between Cpk & Ppk*

Cpk is for short term, Ppk is for long term.

Ppk produces an index number (Cpk=1.33) for the process variation. Cpk references the variation to your specification limits. If you just want to know how much variation the process exhibits, a Ppk measurement is fine. If you want to know how that variation will affect the ability of your process to meet customer requirements (CTQ's), you should use Cpk. It could be argued that the use of Ppk and Cpk (with sufficient sample size) are far more valid estimates of long and short term capability of processes since the 1.5 sigma shift has a shaky statistical foundation.

Cpk tells you what the process is capable of doing in future, assuming it remains in a state of statistical control. Ppk tells you how the process has performed in the past. You cannot use it to predict the future, like with Cpk, because the process is not in a state of control. The values for Cpk and Ppk will converge to almost the same value when the process is in statistical control. That is because

Sigma and the sample standard deviation will be identical (at least as can be distinguished by an F-test). When out of control, the values will be distinctly different, perhaps by a very wide margin.

"Cp and Cpk are for computing the index with respect to the sub grouping of the data (different shifts, machines, operators, etc.), while Pp and Ppk are for the whole process (no sub grouping). For both Ppk and Cpk the 'k' stands for 'centralizing factor'- it assumes the index takes into consideration the fact that the data is maybe not centered (and hence, the index shall be smaller). It is more realistic to use Pp & Ppk than Cp or Cpk as the process variation cannot be tempered with by inappropriate sub grouping. However, Cp and Cpk can be very useful in order to know if, under the best conditions, the process is capable of fitting into the specs or not. It basically gives the best case scenario for the existing process". Cp should always be greater than 2.0 for a good process which is under statistical control. For a good process under statistical control, Cpk should be greater than 1.5

2.3.1.4 *Calculating Cpk and Ppk*

2.3.1.5

$$Cp = \frac{(USL - LSL)}{6*\sigma}$$

$$Cp \text{ --lower limit} = \frac{(Mean - LSL)}{3*\sigma}$$

$$Cp - \text{upper limit} = \frac{(USL - Mean)}{3*\sigma}$$

$$Cpk = Min(Cpl, Cpu)$$

As can be interpreted from the above equations, Cp gives an indication of how narrow the data distribution is relative to the width of the specification limits. Essentially, it indicates how well the process would be able to stay within the specified limits if the data were perfectly centered between those limits.

Cpk compares the widest half of the data distribution to the appropriate specification limit. It indicates whether the process is capable of meeting the specification as indicated by the "worst half" of the measurements. Unlike Cp, the Cpk index measures process capability without assuming the data is well centered. Table 2 shows the formula for CP, Cpk, Pp and Ppk.

Process Capability Indices for Bi-lateral Tolerances					
Index	Variabe Estimate Used	Formula	Index	VEU	Formula
Cp	$\hat{\sigma}$	$C_p = \dfrac{USL - LSL}{6\hat{\sigma}}$	Pp	s	$P_p = \dfrac{USL - LSL}{6s}$
Cpk	$\hat{\sigma}$	$C_{pk} = \min\left[\dfrac{USL-\overline{\overline{X}}}{3\hat{\sigma}}, \dfrac{\overline{\overline{X}}-LSL}{3\hat{\sigma}}\right]$	Ppk	s	$P_{pk} = \min\left[\dfrac{USL-\overline{\overline{X}}}{3s}, \dfrac{\overline{\overline{X}}}{}\right.$
Cr	$\hat{\sigma}$	$C_r = \dfrac{6\hat{\sigma}}{USL-LSL}$	Pr	s	$P_r = \dfrac{6s}{USL-LSL}$
Cpm	$\hat{\sigma}$	$C_{pm} = \dfrac{USL-LSL}{6\sqrt{\hat{\sigma}^2 + \left(\overline{\overline{X}}-TAR\right)^2}}$	Ppm	s	$P_{pm} = \dfrac{USL-LSL}{6\sqrt{s^2 + \left(\overline{\overline{X}}-TAR\right)}}$

"Table 2" process capability index

2.3.1.6 *Process Variation:*

The Six Sigma methodology also focuses on variation, not just averages. As organizations, we tend to focus on how well things work on average, but customers tend to focus on the variation. Consider on-time arrival of airline flights, the airlines measure the percent of on time arrivals. It is not the flights that arrive on time or within a few minutes of on time that are of concern to customers, it is the flights that are late, especially those that are very late (the definition of very late may vary among passengers). Most passengers could not tell you the percent of flights that they have taken that arrived on time, but they can all tell you stories of the late and very late arrivals they have had to deal with. Customers focus on the variation of the process, so as suppliers we must also focus on reducing variation, not just managing the averages.

Six Sigma focuses on the premise that the outcome of any process is a function of the inputs to the process and what happens during the process, stated as $Y = f(x)$. Y is the outcome of the process, generally cycle time or an opportunity for a defect. F is the transform function of the process, the transformation of the inputs into the outcomes. X's are the inputs to the process; typically the sources of variation in the process that need to be controlled to ensure the outcome of the process will consistently meet the customers' requirements. The goal of Six Sigma is the find the vital few X's that must be controlled or changes to the transform function (f) to ensure that Y, the outcome, meets the customers' requirements predictably. Measure phase identifies potential sources of variation, X's.

As mentioned, measuring the capability of the process is required to meet the customers' specifications. There are two ways we assess the process, depending on the type of data. Some processes have attribute data, the outcome is good or bad, such as on time or not on time. When the process has attribute data, the defect is counted by looking at the proportion of bad outcomes to total outcomes. The date will be normalized as though there were one million outcomes (defects per million opportunities) and use a table to determine the sigma level for the process. The process will be declared if it is operating at a Six Sigma level if only 3.4 of the outcomes were bad and 999,996.6 of the outcomes were good of one million opportunities. If, on the other hand, we have continuous data, such as minutes late, then the mean and standard deviation can be calculated for the process. We check to see if the mean for the process is in alignment with the customers' target, if not we may need to recenter the process. We also check the variation; the sigma level of a process is the number of standard deviations that can fit between the mean and the nearest customer specification.

A Six Sigma process has so little variation that six standard deviations can fit between the mean and each specification limit, in other words very little variation. If six standard deviations fit between the mean and the specification limit, and then 99.99966% of the outcomes will be within the customers' specification limits. If it is possible and appropriate to use continuous data, instead of attribute, we would prefer to work with the continuous data because we can gather more information from smaller samples of data.

Historically, we have measured the yield of our processes, by measuring the number of products or services that meet customers' requirements at the end of the process. Studies show that there is no correlation between improvements in the final yield of most processes and the profitability of organization. The final yield of the process does not take rework (the hidden factory) in the process into account. We need to measure our ability to do things right the first time.

In the Six Sigma methodology, we define a defect as an outcome that does not meet the customers' requirements in any one aspect. We measure defects by looking at how many outcomes do not meet the customers' requirements. If we produce 6 items and each item has 5 opportunities (things that can be wrong) for a defect, we can not just consider the number of items produced to get a clear understanding of the performance of the process, 2 items of 6 is a 33% yield for the process. We need to look at each of the 30 opportunities (5 opportunities times 6 items) and the ability of each of those opportunities to be performed correctly the first time, $(16/30)5 = 4.3\%$ yield. An item with multiple defects is more costly to produce and rework, so that it can be delivered to the customer, than an item with no defects. The more defects that occur in any item, the more costly that item is to produce, to meet customers' requirements, because of the amount of rework to the item. Opportunities for a defect should be defined from the customers' perspective, for example a line item on an order or a statement of account activity.

Six Sigma looks at the Rolled Throughput Yield (RTY) of the process, or the ability of the process to produce the outcome correctly the first time. To calculate the Rolled Throughput Yield of the process, we look at the probability that each opportunity for a defect (part of the product or step) in the process will be performed correctly the first time, and then we multiply the first pass yield of each of the steps together to get the Rolled Throughput Yield of the process. If it is difficult to measure the final yield of each step, it is possible to estimate the Rolled Throughput Yield of the process by calculating the average total number of defects that occur when producing an item or providing a service and applying the Poisson distribution, e^{-dpu}, where e (a constant, approximately 2.718) is raised to the negative defects per item or unit.

This gives us the probability of producing an item correctly the first time through the entire process, the cheapest way to meet customers' requirements. This method for measuring yield does correlate to profitability and is used in the Six Sigma methodology.

2.3.2 Process Maps

A process is defined as "a series of steps which convert one or more inputs into one or more outputs." A Process Map is a graphical representation of those steps. The map traces the flow of a physical product (if there is one) and the flow of information through the steps of a process. If there is no physical product, the map is used to document the step-by-step activities involved in providing a service. Process maps may be used to describe a process occurring at any level of the organization. Additionally, they may be expanded or contracted to depict the desired level of detail.

Unlike Relationship Maps, Process Maps depict:

- Accomplishments (actions)
- Decisions
- Sequence of events

There are basically two types of Process Maps:

2.3.2.1   *Linear Process Maps*

Tracks accomplishments, inputs/outputs and decisions as they occur sequentially in the process. Time is depicted on the map as moving generally from left to right and from top to bottom. Typically, time is not represented to scale.

2.3.2.2   *Cross-Functional Process Maps*

While both types of maps are used to describe how work is accomplished, the Cross-Functional Map also shows who (which Functions) is involved in getting it done.

2.3.2.3   *Why Use Process Maps?*

Process Map can:

- Help make an existing process "visible" so that it can be understood more readily by those working in it and managing it.
- Serve as a basis for analysis of a process—to identify aspects of the process that need to be changed as well as those that should be retained.
- Make visible a team's "vision" for improving an existing process or creating a new one.
- Provide a useful framework for determining where to establish measurement points for ongoing management of a process.

2.3.2.4   *How to Create Linear Process Maps*

A Linear Process Map is a type of flowchart. Most processes can be mapped (linearly) using three elements. Process Step (box), Decision (diamond), and Inputs/Outputs (lines).

2.3.2.5 *Function Relationship Maps*

A Function Relationship Map offers a high-level view of an organization. It is a picture of the key relationships, expressed as inputs or outputs, between the organizations internal Functions (departments), and its customers and suppliers. These three types of Business Maps comprise a comprehensive set of tools for use during the definition, analysis, and design phases of the projects. They can be used together or individually, the maps can significantly improve understanding of the system in which an individual works, and subsequently improve that system through enhanced efficiency.

2.3.3 Cause and Effect Diagram

The cause and effect diagram is also called the fishbone chart because of its appearance and the Ishikawa chart after the man who popularized its use in Japan. The lines coming off the core horizontal line are the main causes and the lines coming off those are sub causes. Following are some of the applications of the fishbone diagram.

- Focus attention on one specific issue or problem.
- Organize and display graphically the various theories about what the Root Causes of a problem may be.
- Show the relationship of various factors influencing a problem.
- Cause-and -effect diagrams do not have a statistical basis, but are excellent aids for problem solving.
- Reveal important relationships among various variables and possible causes.
- Provide additional insight into process behaviors.
- Focus the team on the causes, not the symptoms.

2.3.3.1 *Create a Cause-Effect Diagram*

1. Clearly identify and define the problem, symptom, or effect for which the causes must be identified.

2. Place the problem or symptom being explored at the right, enclosed in a box.

3. Draw the central spine as a thick line pointing to it from the left.

4. Brainstorm (or construct an Affinity Diagram) to identify the "major categories" of possible causes (not less than 2 and normally not more than 6 or 7). If other applicable data such as check sheets are present, incorporate them as well.

   You may summarize causes under categories such as:

   - Methods, Machines, Materials, People and Environment

   Place each of the identified "major categories" of causes in a box or on the diagram and connect it to the central spine by a line at an angle of about 70 degrees from the horizontal.

CAUSES

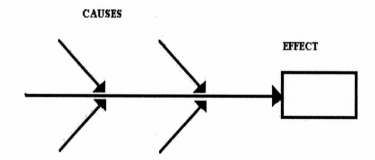

5. Within each "major category" ask, "Why does this condition exist?"

6. Continue to add clauses to each branch until the fishbone is completed.

7. Once all the bones have been completed, identify the likely, actionable Root Cause(s).

2.3.3.2 *Some points to keep in mind*

- State causes, not solutions.

- Take note of causes that appear repeatedly.

- Review each major cause category. Circle the most likely causes on the diagram.

- Test the most likely cause and verify with data.

2.3.4 Measurement system analyses

One of the biggest advantages of the Six Sigma program is the emphasize on the data and measurement and metrics. As much as the organization expands the decisions will be made based on the historical experiences. Implementation of the Six Sigma program forces the managements to make decisions based on the facts and figures and the behavior of the data. As we will explain further the statistical portion of the Six Sigma needs to be high lighted and promoted. Most organizations implement measurement and metrics to develop processes and products to assist in performance improvement and/or benchmarking. In order to ensure that the metrics being collected are relevant, the and Performance Indicators need to be first established. The key performance indicators may have already been established, however they need to be identified and so that the metrics collection and reporting practices are aligned to their needs. In order to establish a measurement system that works robustly for an organization, some preparation is required.

The culture of the organization needs to be evaluated to understand the types of the measurements that required.

- Identify the key performance indicators

33

- Define the metrics and the baseline for measurement

- Establish a project plan to implement the measurement methodologies

- Involve all the department in development of the metrics for the organization

- Use the quality function deployment (house of quality) to prioritize and select the most important metrics which directly impact the customer satisfaction

2.3.5 Graphical Techniques

Graphical techniques are the methods of illustrating the data/ information on a graphical form. Some of the graphical techniques that have been used widely are included:

Bar chart, histogram, line graph, frequency polygon, and stem and leaf chart. The most important part of the charts and graph are the interpretation of the data at a glance. In addition graphs represent quantitative values qualitatively, i.e. by showing numerical values non- numerically.

2.3.5.1 *Bar Chart:*

As the name implies, a bar chart depicts a numerical value using a bar or column of a length related to the value being represented. One bar chart may contain many bars of different lengths allowing direct comparison. Some bar charts are graphically presented in different colors or shadings to assign related but separate values to a common variable, such as smoking rates among males and females in a given country. Such bar charts are called combined bar charts and include a legend that explains the assigned color or shading. Bar charts are appropriate in situations where continuous numerical values are assigned to non-continuous quantities. Bar charts can also be used to directly represent single statistics such as mean, percentage, other statistical values, or even frequencies.

The strength of a bar chart lies in its visual accessibility. The length of the columns can be grasped and compared at a glance. The weakness, however, lies in the same place: it is fairly easy to create misleading bar graphs. This is done primarily by manipulating the scale and origin of the numerical column axis as described in the following example. An incumbent candidate for election to public office may choose to represent a 5% drop in unemployment during his term in office in a more favorable way by using a bar graph that only depicts the upper few percent of the data, i.e. making it look like there was a huge drop in unemployment. By contrast the opponent would choose to use the full 100% scale in order to make it seem like the incumbent's policies made no difference.

2.3.5.2 *Line Graph*

A Line graph is a graphical representation of two independent continuous sets of data. Data values are assigned ordered pairs which are plotted on a pair of axis relating to the variables of the data. These plotted points are then connected by straight lines to form the line graph. Some line graphs may contain more than one line represented by different colors allowing direct comparison of more than one variable in one graph. Such graphs are called combined line graphs and contain a legend to explain the colors. A line graph is appropriate to use when graphing two continuous sets of variables, for example, when graphing the distance an object has traveled relative to the time it took. Both distance and time are continuous variables, therefore it is appropriate to connect the points of the graphs with a straight line. A line graph should be used when the development of a variable is to be revealed. The strength of the line graph is its ability to show trends.

2.3.5.3 *Histogram*

A histogram appears similar to a bar chart, but with columns whose length represents the frequency of a numerical value. A histogram relates these values to a group or class of values out of a continuous set of values, compared to the non-continuous values of the bar graph. This means that the width of each column is the same and represents one class interval. Class intervals are represented by the mid-point at the center of each column. In a histogram, unlike a bar chart, all intervals are represented, even if empty.

Histograms are used to represent the frequency of values in a set of data. The data must be divided into appropriate classes and a frequency table, showing the frequency of occurrence of each class must be made. The height and area of each column in the histogram are proportional to the class frequency.

As with the bar graph the strength of a histogram lies in its visual accessibility. The frequency of classes can be directly assessed without consulting numbers. As a means of comparing class frequencies the histogram is well suited, however, it does suffer the weakness of simplifying all the values in one class into one single value.

2.3.5.4 *Frequency Polygon*

A frequency polygon results when the mid-points of the columns of a histogram are connected by straight lines without drawing the rectangular columns of the histogram. For each class of the data the frequency is established and a dot placed at the correct height on the mid-line of the class, and the resulting dots are connected to form a polygon. The frequency polygon is used to represent the frequency of values in a set of data, just like a histogram. Frequency polygons are commonly used where a more curve-like appearance is desired for frequency representation, while a histogram is used where a more block-like appearance is desired.

Since the vertices of the graph are connected to form a polygon the impression of a smooth transition between classes is created. This can be used to estimate values between classes, clearly an advantage of the frequency polygon. At the same time, this may be misleading if in fact the data is structured such that the values in a given class are all close together and the true picture is more step-like.

2.3.5.5 *Stem and Leaf Chart*

A stem and leaf chart is a device used to sort data efficiently by hand. The data is pre-sorted by its tens- or hundreds- (or bigger) value digit and then sorted a second time by smaller value digits. The tens- or greater digits are written in order in a vertical column (the stem) and the other digits of each value are written in a horizontal line beside each stem value (the leaves). In the second round of sorting the leaves beside each stem are placed in numerical order. The result is a rectangular chart, to be read horizontally, consisting of all the values of a set of data placed in numerical order. A stem and leaf plot is used to sort numerical data by hand. The stem and leaf chart provides an efficient convenient method of sorting large and small quantities of data with minimal likelihood of error. The chart provides a basis for easy tabulation of classes of data. Nowadays the use of stem and leaf charts has been replaced by computerized data management systems.

2.3.6 Measure Phase Checklist

Measure	How? Tools	Deliverables
Select Project Y/ CTQ characteristic to be improved	Customer input, QFD, FMEA, process map, Focused Project Definition, attribute and variable data	Confirmed Project Y and Operational definition. Selected CTQ to improve
Define Performance Standards for the Y	Customer, Blueprints, Existing process documentation, Historic Business Goals and Objectives	Performance Standards (spec limits, target, and defect definition) for Project Y
Measurement System Analysis	Continuous Gage R&R, Test/Retest	Validated Measurement system for project Y
Identify segmentation factors for data collection plan	Segmentation tree, process maps, organizational charts, system diagrams	Identified Segmentation factors
Data collection	Operational definitions and procedures, data collection plan, sampling plan, segmentation factors, detailed process mapping, surveying	Data for Project Y with segmentation factors
Describe and display variation of current performance	Descriptive statistics, box plots, scatter diagram, histogram, pareto, run chart, normality test, segmentation, FMEA, detailed process map, Excel, Minitab, defects, costs, capacity/time constraints	Display variation for project Y data, (Baseline of current process performance)
Containment Plan		Containment plan implemented if needed

2.4 *Analyses Phase*

Analyses Phase involves identifying the gap between the current and desired performance, prioritizing problems and identifying root causes of problems. During the Analyze phase, we search for the few factors, key process inputs that determine the outcome of the process, key process outputs. Using graphical and statistical tools on actual data from the process, we determine which factors drive the outcome of the process and which factors have insufficient ability to affect the outcome. Data sample sizes are chosen to allow us to make decisions at an appropriate confidence level. In this phase, we narrow the 10 – 15 variables that we gathered data on in the Measure phase down to 8 – 10 key input variables. At this stage, we may also determine that we have not gathered data on a sufficient number of input factors, x's, if the current set of factors does not show significance.

Table 3 shows the brief functions and variables for any given process.

$Y = f(X1,X2,X3,.....)$	
Dependent	*Independent*
Output	*Input-Process*
Effect	*Cause*
Symptom	*Problem*
Monitor	*Control*

"Table 3" variable and function example

Tools that are typically used in the Analyze phase include:

- Hypothesis Tests
- Correlation Studies
- Regression Analyses

- Analyses of means and variances
- Mutli-vari studies

Benchmarking the process outputs, products or services, against recognized benchmark standards of performance may also be carried out. During the Analyze phase, we search for the few factors, key process inputs that determine the outcome of the process, key process outputs. Using graphical and statistical tools on actual data from the process, we determine which factors drive the outcome of the process and which factors have insufficient ability to affect the outcome. Data sample sizes are chosen to allow us to make decisions at an appropriate confidence level. At this stage, we may also determine that we have not gathered data on a sufficient number of input factors, X's, if the current set of factors does not show significance.

2.4.1 Correlation Studies

The correlation coefficient a concept from statistics is a measure of how well trends in the predicted values follow trends in the actual values in the past. It is a measure of how well the predicted values from a forecast model "fit" with the real-life data.

The correlation coefficient is a number between 0 and 1. If there is no relationship between the predicted values and the actual values the correlation coefficient is 0 or very low (the predicted values are no better than random numbers). As the strength of the relationship between the predicted values and actual values increases so does the correlation coefficient. A perfect fit gives a coefficient of 1.0. Thus the higher the correlation coefficient the better.

2.4.1.1 *Interpret the Correlation Coefficient*

The following general categories indicate a quick way of interpreting a calculated r value:

0.0 to 0.2 Very weak to negligible correlation

0.2 to 0.4 Weak, low correlation (not very significant)

0.4 to 0.7 Moderate correlation

0.7 to 0.9 Strong, high correlation

0.9 to 1.0 Very strong correlation

2.4.1.2 *The correlation coefficient is then defined as*

$$r = \frac{\sigma_{xy}}{\sigma_x\, \sigma_y}$$

2.4.1.3 *Pearson Correlation Coefficient*

The Pearson Correlation Coefficient is a measure of linear association between 2 variables. It is useful when attempting to determine if there is a significant relationship between these two variables. Values of the correlation coefficient (r) range from -1 to +1. The absolute value of the correlation coefficient indicates the strength of the linear relationship between the variables, with larger absolute values indicating stronger relationships. The sign of the coefficient indicates the direction of the relationship.

The Pearson Correlation Coefficient is calculated from two variables (X and Y), usually with interval or ratio level data. Each variable is assigned a score based on its distance from the mean and these scores are then cross multiplied for each subject, and then summed. A linear relationship should exist between the variables — verified by plotting the data on a scatter diagram. Pearson r computations are sensitive to extreme values in the data.

2.4.1.4 *Compute Pearson's r*

$$r_{xy} = \frac{n\sum XY - \sum X \sum Y}{\sqrt{\left[n\sum X^2 - \left(\sum X\right)^2\right] * \left[n\sum Y^2 - \left(\sum Y\right)^2\right]}}$$

n = number of paired observations

X = variable A, Y = variable B

2.4.2 • Hypothesis Tests

A hypothesis test is a procedure for determining if an assertion about a characteristic of a population is reasonable. To get started, there are some terms to define and assumptions to make:

- The *null hypothesis* is the original assertion. If, for example the null hypothesis is that the average price of a ball is \$1.15. The notation is H_0: $\mu = 1.15$.

- There are three possibilities for the *alternative hypothesis*. You might only be interested in the result if ball prices were actually higher. In this case, the alternative hypothesis is H_1: $\mu > 1.15$. The other possibilities are H_1: $\mu < 1.15$ and H_1: $\mu \neq 1.15$.

- The *significance level* is related to the degree of certainty you require in order to reject the null hypothesis in favor of the alternative. By taking a small sample you cannot be certain about your conclusion. So you decide in advance to reject the null hypothesis if the probability of observing your sampled result is less than the significance level. For a typical significance level of 5%, the notation is $\alpha = 0.05$. For this significance level, the probability of incorrectly rejecting the null hypothesis when it is actually true is 5%. If you need more protection from this error, then choose a lower value of α.

- The *p-value* is the probability of observing the given sample result under the assumption that the null hypothesis is true. If the p-value is less than α, then you reject the null hypothesis. For example, if $\alpha = 0.05$ and the p-value is 0.03, then you reject the null hypothesis.

- The converse is not true. If the p-value is greater than α, you have insufficient evidence to reject the null hypothesis.

- The outputs for many hypothesis test functions also include *confidence intervals*. Loosely speaking, a confidence interval is a range of values that have a chosen probability of containing the true hypothesized quantity. Suppose, in our example, 1.15 is inside a 95% confidence interval for the mean, μ. That is equivalent to being unable to reject the null hypothesis at a significance level of 0.05. Conversely if the $100(1-\alpha)$ confidence interval does not contain 1.15, then you reject the null hypothesis at the α level of significance.

Decision

State of nature (actual situation)	Decide that there is a problem	Decide that there is no problem
There isn't a problem; the situation is as it should be.	**False alarm risk (alpha) or Type I risk** • The risk of an accident when there isn't one • Risk of convicting an innocent defendant • Quality acceptance sampling; risk of rejecting a good lot • SPC; risk of calling the process out of control when it is in control • Design of experiments (DOE or DOX); risk of concluding that there is a difference between the treatments when there isn't	**100% - alpha** • Chance of acquitting an innocent defendant • Quality acceptance sampling; chance of accepting a good lot • SPC: chance of calling the process in control when it is • DOE: conclude that there is no difference between the treatments when there isn't.
There is a problem; the situation requires adjustment	**Power (gamma)** **A test's ability to detect a real problem, or difference** • Chance of an accident • Chance of convicting a guilty defendant • Quality acceptance sampling; chance of rejecting a bad lot • SPC: chance of calling the process out of control when it is • DOE: chance of detecting a difference between the treatments	**Risk of missing the problem: Type II risk (beta)** • Risk of not seeing the wolf· • Risk of acquitting a guilty defendant· • Quality acceptance sampling; risk of shipping a bad lot· • SPC; risk of calling the process in control when it is out of control • DOE: chance of missing a difference between the treatments

2.4.3.2   *Analysis of variance*

Analysis of variance (ANOVA) is used to uncover the main and interaction effects of categorical independent variables (called "factors") on an interval dependent variable. The new general linear model also supports sorts categorical dependents. A "main effect" is the direct effect of an independent variable on the dependent variable. An "interaction effect" is the joint effect of two or more independent variables on the dependent variable. Regression models cannot handle interaction unless explicit cross product interaction terms are added, ANOVA uncovers interaction effects on a built-in basis. There is also a variant for using interval-level control variables (analysis of covariance, ANCOVA), and for the case of multiple dependents, multiple analysis of variance (MANOVA), and there is a combination of MANOVA and ANCOVA called MANCOVA. The key statistic in ANOVA is the F-test of difference of group means, testing if the means of the groups formed by values of the independent variable (or combinations of values for multiple independent variables) are different enough not to have occurred by chance. If the group means do not differ significantly then it is inferred that the independent variable(s) did not have an effect on the dependent variable. If the F test shows that overall the independent variable(s) is (are) related to the dependent variable, then *multiple comparison tests* of significance are used to explore just which value groups of the independent(s) have the most to do with the relationship.

If the data involve repeated measures of the same variable, as in before-after or matched pairs tests, the F-test is computed differently from the usual between-groups design, but the inference logic is the same. There are also a large variety of other ANOVA designs for special purposes, all with the same logic.

Unlike regression, ANOVA does not assume linear relationships and handles interaction effects automatically. Some of its key assumptions are that the groups formed by the independent variable(s) be relatively equal in size and have similar variances on the dependent variable ("homogeneity of variances"). Like regression, ANOVA is a parametric procedure which assumes multivariate normality (the dependent has a normal distribution for each value category of the independent(s)).

2.4.3.3 *Testing means relates to variance*

ANOVA focuses on F-tests of significance of differences in group means, discussed below. If one has an enumeration rather than a sample, then any difference of means is "real." However, when ANOVA is used for comparing two or more different <u>samples</u>, the real means are unknown. The researcher wants to know if the difference in sample méans is enough to conclude the real means do in fact differ among two or more groups (ex., if support for civil liberties differs among Republicans, Democrats, and Independents). The answer depends on:

1. *the size of the difference between group means, <u>and</u>*

2. *the sample sizes in each group. Larger sample sizes give more reliable information and even small differences in means may be significant if the sample sizes are large enough. <u>and</u>*

3. *the variances of the dependent variable. For the same absolute difference in means, the difference is more significant if in each group the civil liberties scores tightly cluster about their respective different means. Likewise, if the civil liberties scores are widely dispersed (have high variance) in each group, then the given difference of means is less significant. The formulas for the t-test (a special case of one-way ANOVA), and for the F-test used in ANOVA, thus reflect three things: the difference in means, group sample sizes, and the group variances. That is, the ANOVA F-test is a function of the variance of the set of group means, the overall mean of all observations, and the variances of the observations in each group weighted for group sample size.*

2.4.3.4 *One-way ANOVA*

tests differences in a single interval dependent variable among two, three, or more groups formed by the categories of a single categorical independent variable. Also known as univariate ANOVA, simple ANOVA, single classification ANOVA, or one-factor ANOVA, this design deals with one independent variable and one dependent variable. It tests whether the groups formed by the categories of the independent variable seem similar (specifically that they have the same pattern of dispersion as measured by comparing estimates of group variances). If the groups seem different, then it is concluded that the independent variable has an effect on the dependent (ex., if different treatment groups have different health outcomes). One may note also that the significance level of a correlation coefficient for the correlation of an interval variable with a dichotomy will be the same as for a one-way ANOVA on the interval variable using the dichotomy as the only factor. This similarity does not extend to categorical variables with greater than two values. The one way analysis of variance allows us to compare several groups of observations, all of which are independent but possibly with a different mean for each group. A test of great importance is whether or not all the means are equal. The observations all arise from one of several different groups (or have been exposed to one of several different treatments in an experiment). We are classifying 'one-way' according to the group or treatment.

2.4.3.5 *Two Way Analysis of Variance*

Two Way Analysis of Variance is a way of studying the effects of two factors separately (their main effects) and (sometimes) together (their interaction effect). analyzes one interval dependent in terms of the categories (groups) formed by two independents, one of which may be conceived as a control variable. Two-way ANOVA tests whether the groups formed by the categories of the independent variables have similar centroids. Two-way ANOVA is less sensitive than one-way

ANOVA to moderate violations of the assumption of homogeneity of variances across the groups.

2.4.3.6 *Multivariate or n-way ANOVA*

To generalize, n-way ANOVA deals with n independents. It should be noted that as the number of independents increases, the number of potential interactions proliferates. Two independents have a single first-order interaction (AB). Three independents have three first order interactions (AB,AC,BC) and one second-order interaction (ABC), or four in all. Four independents have six first-order (AB,AC,AD,BC,BC,CD), three second-order (ABC, ACD, BCD), and one third-order (ABCD) interaction, or 10 in all. As the number of interactions increase, it becomes increasingly difficult to interpret the model.

2.4.3.7 *Designs*

ANOVA and ANCOVA have a number of different experimental designs. The alternative designs affect how the F-ratio is computed in generating the ANOVA table. However, regardless of design, the ANOVA table is interpreted similarly — the significance of the F-ratio indicates the significance of each main and interaction effect (and each covariate effect in ANCOVA).

2.4.3.8 *Between-groups ANOVA design*

When a dependent variable is measured for independent groups of sample members, where each group is exposed to a different condition, the set of conditions is called a *between-subjects factor*. The groups correspond to conditions, which are categories of a categorical independent variable. For the *experimental mode*, the conditions are assigned randomly to subjects by the researcher, or subjects are assigned randomly to exposure to the conditions, which is equivalent. For the *non-experimental mode*, the conditions are simply measures of the independent variable for each group. For instance, four random groups might all be asked to take a performance test (the interval dependent variable) but each group might be exposed to different levels of noise distraction (the categorical independent variable).

This is the usual ANOVA design. There is one set of subjects: the "groups" refer to the subset of subjects associated with each category of the independent variable (in one-way ANOVA) or with each cell formed by multiple categorical independents (in multivariate ANOVA). After measurements are taken for each group, analysis of variance is computed to see if the variance on the dependent variable between groups is different from the variance within groups. Just by chance, one would expect the variance between groups to be as large as the variance within groups. If the variance between groups is enough larger than the variance within groups, as measured by the F ratio (discussed below), then it is concluded that the grouping factor (the independent variable(s) does/do have a significant effect.

2.4.3.9 *Completely randomized design*

Completely randomized design is simply between-groups ANOVA design for the experimental mode. Randomization is an effort to control for all unmeasured factors. When there is a priori reason for thinking some additional independent variable is important, the additional variable may be controlled explicitly by a block design (see below) if categorical, or by ANCOVA if it is a continuous variable. In the non-experimental mode, where there is no control by randomization, it is all the more important to control explicitly by these methods.

The analysis of variance approach to regression analysis, also called ANOVA is useful for more complex regression models and for other types of statistical models. To understand this approach, several variables must be defined. As previously defined, the variation of the dependent variable from the model is:

$$y_1 - ybar$$

From variation, the total sum of squares (SSTO) can be calculated:

$$SSTO = \Sigma(y_1 - ybar)^2$$

When SSTO is equal to zero, all of the observations are the same. The next variable to calculate is the error sum of squares (SSE):

$$SSE = \Sigma(y_1 - ybar)^2$$

If the SSE is equal to zero, all of the observations are on the regression line, and the higher the SSE, the greater the variation from the line. The difference between the SSTO and SSE is the regression sum of squares (SSR):

SSR = SSTO - SSE   OR   SSTO=SSE+SSR

These sums of squares provide the values for the first column of the ANOVA table, which looks like this:

Source of variation .	SS
Regression	SSR
Error	SSE
Total	SSTO

The next column contains the degrees of freedom (df). SSTO has n-1 degrees of freedom, SSE has n-2 degrees of freedom, and SSR has 1 degree of freedom. The degrees of freedom for SSR and SSE add to the degrees of freedom for SSTO:

n-2=1=n-1

The table now looks like this:

Source of variation	SS	df
Regression	SSR	1
Error	SSE	n-2
Total	SSTO	n-1

The third column contains the means squared (ms) and can be calculated by dividing the sums squared by their degrees of freedom, therefore:

$$MSR = \frac{SSR}{1}$$

and

$$MSR = \frac{SSE}{n-2}$$

Note that the means squared are not additive, and no value for total mean squares appears in the final table:

Source of variation	SS	df	ms
Regression	SSR	1	SSR/1
Error	SSE	n-2	SSE/n-2
Total	SSTO	n-1	

This table provides the values needed to compute the test statistic F*:

$$F^* = \frac{MSR}{MSE}$$

which tests the hypotheses:

$$C1: \beta_2 = 0$$

$$C2: \beta_2 \neq 0$$

Since the F* distribution is described as:

$$F(1-\alpha, 1, n-2)$$

Where $1=v_1$ and $n-2=v_2$ in the F distribution table, the decision rule is:

If $F^* \leq F(1-\alpha, 1, n-2)$, conclude C_1

If $F^* > F(1-\alpha, 1, n-2)$, conclude C_2

2.4.4 Analyze Phase Checklist

Analyze	How?	Deliverables
Process Capability of existing process	Cp, Cpk, basic statistics, graphical analysis, sampling Normality test, DPMO, Sigma Level, Excel, Minitab, Project plan critical paths (for cycle times)	Process Capability for Project Y
Define Improvement Goal	Benchmarking, competitive analysis, business goals	Statistically define Improvement Goal for Project Y
Identify significant segments of data	Hypothesis tests, multi-vari, graphical analysis, pareto	Segmented and stratified data set, significant factors(clues to root cause)
Identify possible X's	Fishbone (Cause and Effect),FMEA, Process inputs, "5 Why's	List of Possible X's, collect additional data with identified X's
Identify critical X's	Hypothesis tests, Sub-process maps, DOE, Paired Comparison DOE-Screening	Validated root cause/ critical X's identified
Quantify the financial benefit	Cost reduction, revenue opportunity	Financial benefits and timing

2.5 *Improve Phase*

Improve Phase involves generating the improvement solutions and fixing problems to prevent them from reoccurring so that the required financial and other performance goals are met. As we move into the Improve phase, we are ready to develop and test potential improvements to the process. We are searching for the optimal process, by systematically experimenting with the key process inputs, X's.

Once we have one or more likely alternatives, we test them prior to implementation to ensure the changes will have the desired effect on the key process outcomes. Tools typically used in the Improve phase include:

2.5.1 Design of Experiments

A strategy for planning research known as *design of experiments* (DOE) was first introduced in the early 1920s when a scientist at a small agricultural research station in England, Sir Ronald Fisher, showed how one could conduct valid experiments in the presence of many naturally fluctuating conditions such as temperature, soil condition, and rainfall. The design principles that he developed for agricultural experiments have been successfully adapted to industrial and military applications since the 1940s. The basic idea is to devise a small set of experiments, in which all pertinent factors are varied systematically. This set usually does not include more than ten to twenty experiments. The subsequent analysis of the resulting experimental data will identify the optimal conditions, the factors that most influence the results and those that do not, the presence of interactions and synergisms, and so on. The most important aspect of design of experiments is that they provide a strict mathematical framework for changing all pertinent factors simultaneously, and achieve this in a small number of experimental runs. Most of us can only grasp the effect of one factor at a time in our minds, and that leads to the inefficient COST approach. We need the mathematics (and the computer) to keep track of the factors and their combinations.

2.5.1.1 *Different designs are used in different situations*

Depending on what you already know about the problem, different classes of designs are suitable for laying out the set of experiments.

- Screening designs are used at the beginning of an investigation. Here the primary objective is to reduce the initial set of factors that may affect the result to a small set of factors that together have the dominating influence.

- Response surface designs are used later in an investigation for the final optimization, and when one wishes to develop a model relating the dominating factors to the response variables that quantify the results (performance, yield, quality).

- Sometimes simplex designs, and steepest ascent approaches are used to achieve optimal conditions in problems where experiments can only be done one at a time in a sequence. They should then be preceded by a screening.

- In addition, special designs are needed in mixture problems. This type of problem is common in the chemical, food and beverage, cosmetics, and drug industries. They arise because here the factors express the percentages of constituents, and add up to 100 percent. This introduces a constraint on the design and must be handled with special tools and models.

2.5.1.2 *Analyzing the resulting experimental data*

After the planning stage when set of experiments are laid out according to a design planned made either in parallel or one another. Each experiment gives results i.e. values response variables. This gives a model relating the factors to the results, showing which factors are important, and how they combine in influencing the results. The model is then used to make predictions, e.g. how to set the factors to achieve desired (optimal) results.

2.5.1.3 *Treatment*

In experiments, a treatment is something that researchers administer to experimental units . For example, a corn field is divided into four, each part is 'treated' with a different fertilizer to see which produces the most corn; a teacher practices different teaching methods on different groups in her class to see which yields the best results; a doctor treats a patient with a skin condition with different creams to see which is most effective.

Treatments are administered to experimental units by 'level', where level implies amount or magnitude. For example, if the experimental units were given 5mg, 10mg, 15mg of a medication, those amounts would be three levels of the treatment. 'Level' is also used for categorical variables, such as Drugs A, B, and C, where the three are different kinds of drug, not different amounts of the same thing.

2.5.1.4 *Factor*

A factor of an experiment is a controlled independent variable; a variable whose levels are set by the experimenter. A factor is a general type or category of treatments. Different treatments constitute different levels of a factor. For example, three different groups of runners are subjected to different training methods. The runners are the experimental units, the training methods, the treatments, where the three types of training methods constitute three levels of the factor 'type of training'.

2.5.2 Piloting Your Solution

During the Improve phase, solutions are proposed, developed, evaluated and implemented. The possible solutions that serve as a starting point are referred to as opportunities. An opportunity is a chance to improve a business process, or some portion of a business process (e.g., activities, deliverables, participants, etc.). Opportunities spell out changes to the business that allows the performance goals to be met to some degree. Opportunities require time, costs, risks, and support to enact changes and attains goals.

Other models that describe the business domain are examined to identify opportunities for improvement. The models to be examined include the Critical Goal Model, Cause and Effect Model, Organization Model, and Workflow Model.

The following are a set of sample questions that can be asked to identify opportunities:

Inspect the Critical Goal Model

- How can the organization reach the goal?

- Where can the process are improved to increase quality?

- What efficiencies can be realized to increase process speed (responsiveness)?

- What inefficiencies can be driven out of the process to reduce cost?

- Inspect the Cause and Effect Model

- What improvement can be implemented to address a problem?

- What opportunity can be realized by addressing the process as a whole?

- What assumptions in the current process design are no longer valid?

- What has changed in the environment that can be leveraged as a competitive advantage?

2.5.2.1 *Inspect the Organization Model*

- Should the organization be restructured around the processes?

- Can the skill sets within the organization be improved to perform the activities?

- Can the organization (or its workers) be relocated to more conveniently pass work and deliverables?

2.5.2.2 *Activity Opportunities:*

- Can the activity be more cost effective (ratio of value to cost)?

- Can the activity be performed more quickly?

- Can the activity make more efficient use of its resources?

- Should the skill level of those performing the opportunity be raised?

- Would different tools or resources improve the performance of the activity?

2.5.2.3 *Supplier Opportunities:*

- Can we communicate with our suppliers more effectively?

- Can we assist in improving our supplier's performance?

- Are there other suppliers capable of providing resources?

- Can the supplier perform some of the business process's initial activities to relieve the business domain of that responsibility?

2.5.2.4 *Resource Opportunities:*

- Are there other resources available of higher quality?

- Can resources be supplied in a form that is more usable by the business process?

- Can resources be acquired in a more timely fashion?

2.5.2.5 *Customer Opportunities*

- Can communication with our customers be more effective?

- Are there other markets for our products?

2.5.2.6 *Implementation Planning*

- Are there customer activities the business domain can perform to better meet the needs of the customer?

- Can the customer perform some of the business domain's final activities, allowing for reduced prices or quicker response?

2.5.2.7 *Product and Service Opportunities*

- Can the quality of our products and services be increased?

- Can products and services be provided to the customer in a more convenient form?

- Intermediate Deliverable Opportunities:

- Can deliverables be passed between activities more quickly?

- Can deliverables be put in a usable form for the receiving activity?

- Can the number of work flows be minimized to reduce hand-offs?

Typically, several opportunities are available to assist the business in meeting its goals. The key is choosing the right combination of opportunities and the right sequence of change implementation to yield the desired results. Costs, time, risks, and support levels must be weighed against the benefits of pursuing the opportunities.

2.5.3 Improve Phase Checklist

Improve	How? Tools	Deliverables
Identify Solution Alternatives to address critical X's	To-be" process map, brainstorm, alternative solutions evaluation, robust design, error-proofing, cost benefit	Proposed Solution; cost benefit analysis
Discover Variable Relationships between X's and Y's	DOE Factorial Designs, Regression, transfer functions between X's and Y	Confirmed Solution, optimal settings and tolerances for X's
Perform trade-offs, optimize settings for the X's	FMEA on solution, Simulation, Tolerance, decision making matrix interactions balanced scorecard	Refined solution
Pilot/ Implement solution	Project planning, evaluate implications of improvement to other business/ customer	Measured improvements on the Y

2.6 Control Phase

Control phase involves implementing the improved process in a way that 'holds the gains'. Standards of operation will be documented in systems such as ISO 9000 and standards of performance will be established using techniques like SPC. After a 'running-in' period, the process capability is calculated again to establish whether the performance gains are being sustained. The cycle is repeated, if further performance shortfalls are identified.

The Control phase targets the on-going management of the processes that were modified during the Improve phase. Unlike the other Six Sigma phases, the

Control phase does not have a completion date, rather it ensures that the problems that were fixed stay fixed.

2.6.1 Realized Objective during the Control Phase

1. Verify the solution works as planned. During the Improve phase, the process needs to be remodeled and simulated to estimate the degree of performance improvement. The improved process is initially implemented in one organizations or locations for a trial period. During this period, a number of measurements are taken to verify that actual performance meets expectations. The improved process should be corrected as necessary based on this 'real world' data.

2. Roll out the solution across the enterprise. Based on the success of the trial Implementation, the improved process is implemented across all portions of the enterprise performing the process. At this point process ownership should be established if it has not already been done so. The process owner or owners are responsible for the on-going management and improvement of the business process.

3. Make sure the problem stays fixed. There is a basic tendency for the process performers to drift back to their previous method of work after a period of time. A subset of the process measurements are left in place to assure continued consistency. The key factor in assuring process consistency is standardization and documentation of the new process.

4. Ensure process flexibility for further improvements. The models can be updated with continuing improvements and made immediately available to anyone who wants to view them, even remotely through a network.

Finally, in the Control phase, we implement the approved improvement. We mistake proof the new process, train the people in the new process, update documentation to reflect the changes, and automate appropriate portions of the process. We also build a control plan to ensure long term sustainability and implement a control mechanism such as statistical process controls. The control plan is tested to ensure that it will allow the process owner to sustain the improvement once the black belt and project team are no longer focused on the process. This is the stage where we lock in the change.

2.6.2 Typical tools used during the Control phase include:

- Control plan

- Training

- ISO 9000 standards as guidelines for work flow standardization, documentation and auditing

- Statistical process control charts

2.6.3 Control Charts:

Control charting is one of the tools of Statistical Quality Control(SQC). It is the most technically sophisticated tool of SQC. Dr. Walter A. Shewhart of the Bell Telephone Labs developed it in the 1920s. Dr. Shewhart developed the control charts as an statistical approach to the study of manufacturing process variation for the purpose of improving the economic effectiveness of the process. These methods are based on continuous monitoring of process variation.

2.6.3.1 *Background Information*

A typical control chart is a graphical display of a quality characteristic that has been measured or computed from a sample versus the sample number or time. The chart contains a center line that represents the average value of the quality characteristic corresponding to the in-control state. Two other horizontal lines, called the upper control limit (UCL) and the lower control limit (LCL) are also drawn. These control limits are chosen so that if the process is in control, nearly all of the sample points will fall between them. As long as the points plot within the control limits, the process is assumed to be in control, and no action is necessary.

However, a point that plots outside of the control limits is interpreted as evidence that the process is out of control, and investigation and corrective action is required to find and eliminate the assignable causes responsible for this behavior. The control points are connected with straight line segments for easy visualization.

Even if all the points plot inside the control limits, if they behave in a systematic or nonrandom manner, then this is an indication that the process is out of control.

2.6.3.2 *Uses of Control charts*

Control chart is a device for describing in a precise manner what is meant by statistical control. Its uses are:

- It is a proven technique for improving productivity.

- It is effective in defect prevention.

- It prevents unnecessary process adjustments.

- It provides diagnostic information.

- It provides information about process capability.

The control chart equations are a bit intimidating, however all statistical software and nearly all spreadsheets can generate control charts. Control charts are a powerful, simple, and graphical tool for managing processes. They are an especially useful tool for dynamic processes such as most clinical processes.

2.6.4 Types of Variation

2.6.4.1 *Common Cause*

Normal random process noise have small effect on the process are inherent to the process because of the nature of the system, the way the system is managed, the way the process is organized and operated can only be removed by making modifications to the process.

2.6.4.2 *Special Cause*

Significant signal of a significant with an assignable cause localized in nature, exceptions to the system considered abnormalities, often specific to a certain operator, certain machine, certain batch of material, etc.

Investigation and removal of variations due to special causes is key to process Improvement.

2.6.5 Types of Data

- Attributes - Characteristic or categorized data rather than the data itself.

- Variables - Numeric data. Variables data is convertible to attribute data but important information is lost.

2.6.6 Control Chart Types

- *Variables :Individuals*

2.6.6.1 *X-bar/R:*

The x Bar and R Charts using data obtained from measurements are the most powerful of all control charts. Two charts that are used together to chart variables data are called $\overline{X}$ and R charts. The sample average is represented by $\overline{X}$ and R is the range. The range is the difference between the highest and lowest number in the sample. The charts are very effective indicators of problems in the process and also indicate when the problems have been cleared. The chart is calculated as follows:

Upper control limit (UCL) = X bar+A_2Rbar

Lower control limit (LCL) = Xbar- A_2 Rbar

Control limits for the R chart are calculated as follows:

Upper control limit (UCL) = D4Rbar

Lower control limit (LCL) = D3Rbar

The factors A_2, D_3, and D_4 are listed in the table of factors.

2.6.6.2 *X-bar/S*

X-bar/S always used when the sample size is greater than 10. The standard deviation should also be used when $1<n<10$. R compared to s: Use X bar and R/s charts

$$UCL = \bar{s} + 3\frac{\bar{s}}{c_4}\sqrt{1 - c_4^2}$$

$$LCL = \bar{s} - 3\frac{\bar{s}}{c_4}\sqrt{1 - c_4^2}$$

- *Attributes*

2.6.6.3 *p chart:*

P Chart is used to determine if the rate of nonconforming product is stable and detect when a deviation from stability has occurred. The argument can be made that a LCL should not exist, since rates of nonconforming product outside the LCL is in fact a good thing; we WANT low rates of nonconforming product. However, if we treat these LCL violations as simply another search for an assignable cause, we may learn for the drop in nonconformities rate and be able to permanently improve the process. p Charts can be used when the subgroups are not of equal size. The np chart is used in the more limited case of equal subgroups

$$UCL = pbar + 3\sqrt{\frac{pbar(1-pbar)}{n(i)}}$$

$$LCL = pbar - 3\sqrt{\frac{pbar(1-pbar)}{n(i)}}$$

2.6.6.4 *np chart:*

The np Chart can be used for the special case when the subgroups are of equal size. Then it is not necessary to convert nonconforming counts into the proportions phat(i). Rather, one can directly plot the counts x(i) versus the subgroup number i.

Steps in Constructing an np Chart.

Determine the size of the subgroups needed. The size, n, has to be sufficiently large to have defects present in the subgroup most of the time. If we have some idea as to what the historical rate of nonconformance, p, is we can use the following formula to estimate the subgroup size: n=3/p. find pbar.

$$pbar = \frac{x(1) + x(2) + \ldots + x(k)}{k \cdot n}$$

Find the UCL and LCL where:

$$UCL = n \cdot pbar + 3\sqrt{n \cdot pbar(1-pbar)}$$
$$LCL = n \cdot pbar - 3\sqrt{n \cdot pbar(1-pbar)}$$

2.6.6.5 *c chart:*

The c Chart measures the number of nonconformities per "unit" and is denoted by c. This "unit" is commonly referred to as an inspection unit and may be "per day" or "per square foot" of some other predetermined sensible rate.

Steps in Constructing a c Chart

Determine cbar.

$$cbar = \frac{1}{k} \sum c(i)$$

There are k inspection units and c(i) is the number of nonconformities in the ith sample.

Since the mean and variance of the underlying Poisson distribution are equal,

$\hat{\sigma}^2 = cbar$ Thus, $\hat{\sigma} = \sqrt{cbar}$

and the UCL and LCL are:

$$UCL = cbar + 3 \cdot \sqrt{cbar}$$
$$LCL = cbar - 3 \cdot \sqrt{cbar}$$

2.6.6.6 u chart:

The u Chart is used when it is not possible to have an inspection unit of a fixed size (e.g., 12 defects counted in one square foot), rather the number of nonconformities is *per inspection unit* where the inspection unit may not be exactly one square foot...it may be an intact panel or other object, different in size than exactly one square foot. When it is converted into a ratio per square foot, or some other measure, it may be controlled with a u chart. Notice that the number no longer has to be integer as with the c chart.

Steps in Constructing a u Chart:

Find the number of nonconformities, c(i) and the number of inspection units, n(i), in each sample i.

Compute u(i)=c(i)/n(i)

Determine the centerline of the u chart:

$$ubar = \frac{total\_nonconformities\_in\_k\_subgroups}{total\_number\_of\_inspection\_units}$$

$$ubar = \frac{c(1) + c(2) + ... + c(k)}{n(1) + n(2) + ... + n(k)}$$

The u chart has individual control limits for each subgroup i.

$$UCL = ubar + 3\sqrt{\frac{ubar}{n(i)}}$$

$$UCL = ubar - 3\sqrt{\frac{ubar}{n(i)}}$$

2.6.6.7 *DPMO Chart:*

DPMO as we have learned is stand for defect per million opportunity. Since the Six Sigma methodologies recommend to measure the process performance by DPM (Defect per million), it will be beneficial to create the control charts to reflect the number of defects per million.

$$DPU\ (defect\ per\ million) = \frac{Number\ of\ defect}{Total\ number\ of\ inspected\ /tested}$$

DPMO= (DPU*1000,000)/ Number of opportunities

$$\overline{DPMO}\ bar = \frac{\Sigma(dpmo)}{N}$$

UCL (Upper control limit) = DPMO + 3000√ $\overline{DPMO\ bar}$ / $\overline{N}$ * Opportunities

2.6.7 Analysis of Patterns on Control Charts

A control chart may indicate an out-of-control condition either when one or more points fall beyond the control limits, or when the plotted points exhibit some nonrandom pattern of behavior. The process is out of control if any one or more of the criteria is met.

- One or more points outside of the control limits. This pattern may indicate:

  A special cause of variance from a material, equipment, method, or measurement system change. Mismeasurement of a part or parts. Miscalculated or misplotted data points.

  Miscalculated or misplotted control limits.

- A run of eight points on one side of the center line. This pattern indicates a shift in the process output from changes in the equipment, methods, or materials or a shift in the measurement system.

- Two of three consecutive points outside the 2-sigma warning limits but still inside the control limits. This may be the result of a large shift in the process in the equipment, methods, materials, or operator or a shift in the measurement system.

- Four of five consecutive points beyond the 1-sigma limits.

  An unusual or nonrandom pattern in the data.

- A trend of seven points in a row upward or downward. This may show Gradual deterioration or wear in equipment.

  Cycling of data can indicate:

- Temperature or other recurring changes in the environment.

- Differences between operators or operator techniques.

- Regular rotation of machines.

- Differences in measuring or testing devices that are being used in order.

- Several points near a warning or control limit.

2.6.8 Control Phase Checklist

Control	How? Tools	Deliverables
Define & Validate Measurement System on X's in Actual Application	Continuous Gage R&R, Test/Retest, Attribute R&R	Validated Measurement System
Determine Improved Process Capability	Cp, Cpk, DPMO, Sigma Level, measured variation	Statistical confirmation improvement goal is realized Post-Improvement Process Capability
Implement Process Control	Control Charts, Hypothesis tests, Mistake Proof, FMEA, monitoring, response plan, process compliance audits/ measurements, Org Change Management, stakeholder analysis	Process Control Plan/ Owner Signoff, Sustained Solution, Documentation, control measurements and owners
Close project		Validated cost/benefit analysis; project closure communication

2.7 1.5 Sigma Shift

The magnitude of the shift may vary, but empirical evidence indicates that 1.5 is about average. Does this shift exist in the software process? While it will take time to build sufficient data repositories to verify this assumption within the software and systems sector, it is reasonable to presume that there are factors that would contribute to such a shift.

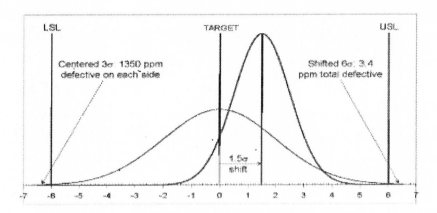

"Figure 4" 1.5 sigma shift capability

It is a complex subject in Six Sigma, and traditionally includes a value of 1.5 sigma short to long term shift in the calculations. The shift is the degradation experienced between the short-term best process capability and the long-term process performance when all possible process time-related causes of variation have been added in. From work undertaken at Motorola an empirical figure of 1.5 sigma is now taken as 'standard' for the worst case shift, although some today now advocate that this figure could be lower.

In theory: Process Sigma (short term) = 1.5 + Process Sigma (long term)

Short term: is the process capability without any time related variation, for example one shift, one operator, one material batch, one machine setting and so on. This is often taken as the best capability figure, and by convention the Process Sigma value of every process is stated as the best short-term value.

Long term: is the process capability with every possible time related variation included, for example every shift, every operator, every material batch and supplier, every machine setting and so on. This is often taken as the worst capability figure, and by convention the short-term Process Sigma is related to the DPMO at the long-term state, typically shifted by 1.5 sigma from the short term state.

If your data is *short term*, then the DPMO figure can be related back to (short term) process sigma without any shift, however the customer will experience the DPMO defect rate for the long term process state.

If your data is *long term*, then the DPMO figure must have the 1.5 sigma shift added in to return to the short term Process Sigma value. Transactional process data is typically long term. How this calculator works:

The sample and population sizes are used to extrapolate the *total defects* in the population. The population and defect opportunity are used to calculate the *total number of opportunities*. DPMO is then calculated using:

DPMO = (total defects * 1,000,000) / (total opportunities* total unit tested)

The DPMO is then converted to the number of standard deviations for the equivalent right-hand tail fraction of the normal distribution. This number of standard deviations is the base process sigma value. The base process sigma is then adjusted to obtain the *long term process sigma value* - if the data is long term then no adjustment is needed, however for short term data the assumed short-long term shift is added in. This is 1.5 sigma by default, but can be adjusted to any value of choice. Intermediate states between short and long term apply this shift in proportion. The (official) short term process sigma value is then calculated by adding back a standard 1.5 sigma shift.

For example - our data is very short term, we anticipate a maximum 1 sigma shift between short and long term, and the DPMO works out to 6210. This is equivalent to the right tail of a normal distribution at 2.5 sigma (short term). We subtract 1 sigma to get our anticipated 1.5 sigma (long term) [we expect that over the longest term, the customer would experience 66800 DPMO], and then add back the standard 1.5 sigma to arrive at the (official) Process Sigma value of 3.0, Where the process shift is not known the default 1.5 shift should be used. Since 6 sigma = 3.4 DPMO as standard, the 1.5 correction factor will always return any long term DPMO back to an 'official' short term Process Sigma Value.

3 CHAPTER THREE

3.1 *Design for Six Sigma (DFSS)*

DFSS shares with DMAIC (Define-Measure-Analyze-Improve-Control) the same emphasis on customer satisfaction and critical-to-quality factors; however, instead of analyzing existing processes, the focus is to create new products and services with built-in Six Sigma quality. These products and services are efficient, high-yielding, and robust to process variations. A resource for DFSS is the asset library of processes, product components and practices. Modeling, simulation, lean, systems thinking and robust design techniques are among the analytical tools that may be useful in DFSS. DFSS, as applied to the redesign of existing processes and products, is believed to be a critical step in moving from five-sigma to Six Sigma performance.

DFSS is an established data-driven methodology based on analytical tools, with the ability to prevent and predict defects in the design of a product, service, or process.

It is part of a broad range of Six Sigma processes that identify and improve efficiency and quality in virtually everything an organization does throughout its worldwide operations.

Six Sigma has provided significant benefits to many companies, such as General Electric, Motorola, Honeywell, Samsung Electronics of South Korea, and Telefonica in Spain. It is helping these organizations and many others control costs; improve quality in products, manufacturing, and transactional processes (internally as well as for customer services); and increase the efficiency and productivity of complete supply chains. In short, Six Sigma is financially rewarding—with ROI ranging from 10:1 to over 100:1.

Design for Six Sigma (DFSS) focuses on developing products and services that deliver critical customer requirements at Six Sigma levels of performance. One tangible benefit of DFSS is delighting customers through superior designs.

Delighting customers demands a comprehensive understanding of both customer needs and organizational capabilities.

3.2 Design for Six Sigma Method

Design for Six Sigma utilizes the most powerful tools and methods known today for developing optimized designs. These tools and methods are ready to plug directly into your current product development process.

DFSS provides many tangible benefits to companies. For instance, the DFSS approach results in long term cost reductions associated with a product. There are many ways this savings is realized; the most pronounced are:

development and verification,

(2) manufacturing (both tooling & facilities and variable), and

(3) after sale service and support. DFSS reduces time to and thereby ensures fresher product feature sets. DFSS also improves quality at introduction through a clearer definition of customer priorities and effective incorporation of manufacturing issues.

Reducing life cycle costs is one of the primary goals of a Design for Six Sigma project. In the product development process, when first applying DFSS, some costs will go up while others will go down. The increase in initial cost is associated with capturing and understanding customer requirements; however, this cost will easily be recovered through improvement of the downstream activities. Design and development activities will have clear priorities and requirement sets thereby reducing design iterations and clarifying verification approaches. Manufacturing issues are considered in every DFSS project thereby maximizing re-use of facilities and equipment. Where investment is required, the optimum configuration and capability can be established with clear data on requirements.

Service and support issues will be addressed throughout the DFSS project so that the two major cost drivers (insufficient function and failures) are alleviated prior to release of the product.

Time to develop new products is a critical success factor in almost any business today. DFSS facilitates long term time reduction by deploying lessons learned throughout the development and manufacturing set-up process. First DFSS incorporates a customer knowledge base through. Next DFSS incorporates logical, objective based tradeoffs for time to launch verses customer/market benefit. Where time is spent, it will be clearly value added.

Quality, it turns out, is the key outcome from all of the prior mentioned activities. Being able to improve quality, and at the same time improve cost and cycle time, is the name of the game. The DFSS approach is very strong here:

establish clear requirement sets (customer, company, and regulatory)

look for inventive ways to satisfy those requirements

mitigate the things that could go wrong,

optimize the function of the design,

optimize the cost/benefit associated with the manufacturing tolerances, and

verify that product meets the requirement set. The tools and methods built into DFSS are the key enablers for accomplishing all of the goals.

Putting together and managing a DFSS project is not a trivial matter. All of the key enablers must be in place to realize the maximum benefit. All of the players need to know what role they play and their performance must be tied to their rewards.

People that have a significant part in the DFSS process need to be properly trained in the tools and methods that they will be expected to utilize or support. To that end, a formal certification process is established to insure that all critical skills and knowledge are in place to support a successful DFSS project.

A proper DFSS team represents all of the key functions that contribute to defining the product (and its constraints), designing the product, testing the product, manufacturing the product and servicing the product. Some of the representatives will see the project from the beginning to the end; others will contribute only during specific phases.

Proper representation by function is important; if the selected representative does not have the requisite functional expertise and knowledge their contribution will be negligible (and sometimes negative). Each team member has specific duties and responsibilities that must be effectively carried out to make the project successful.

Supporting a DFSS project requires many levels of involvement. First are the company sponsors. They need to allocate resources, interface to the companies governing body, and ensure that the project is on track.

The DFSS Expert helps the teams and the Project Leaders with the more difficult aspects of a DFSS project; they act more as consultants than managers. They must be experts in the tools and methods of DFSS as well as have good consulting skills.

The DFSS Project Leader manages the day-to-day activities of the team – they are the integrator of all of the various tools and methods into a successful product.

DFSS Practitioners bring specific skills and knowledge to the team, some are core team members, and others only support as needed. Examples include testing, finance, manufacturing, etc...

No DFSS project will be successful without a strong sponsor. This person must have the authority to make significant things happen, anything from spending money to redirecting company resources must be within their purview. The team cannot be expected to fight for support at every step of the process and still have time to do their primary job, which is to bring a Six Sigma product to market.

The sponsor is generally the executive that would have been responsible for the new product, but here their primary role is to support the process.

The DFSS Expert is the primary source for technical knowledge around the Six Sigma tools and methods. As such, this person must not only be highly respected technically, but must also be able to effectively consult with teams and senior management. This person takes years to develop the proper knowledge base and a substantial part of that is company specific.

This is not a typical technical position, for instance, if the DFSS Expert is subordinate to the project Sponsor, you will not get the type of reviews and challenges that the sponsor and team need to stay on track.

The certification process for a DFSS Expert is substantial; it cannot be completed in less than one year, even if the person has all of the required knowledge in the tools. Expert certification attests to more than knowledge and skills; it requires significant successes, consulting on a number of projects.

A DFSS Project Leader is responsible for coordinating the various tasks and activities that contribute to a successful project. Many of these tasks and activities are defined and deployed as part of a company's current new product introduction process. The key challenge for a DFSS Project Leader is effectively integrating the DFSS tools and methods into the current (NPI) process.

The Project Leader may at times need help integrating the DFSS tools and methods into their project; this is where the DFSS Expert comes in, they provide the guidance and technical support to help keep the Project Leader on Track.

There are several phases for a DFSS leader to take in order to accomplish a project.

In each phase the DFSS expert should use specific tools and methods and puts them in the context of the full scale project.

Phase I: Start a new product introduction process

Timing for Phases II, III & IV is more flexible as their concepts play out in an iterative fashion throughout the new product introduction process.

Phase I has two primary objectives, (1) get the project started on the right foot, and (2) clearly define the product requirements to design to. Each of the deliverables plays a part in meeting these objectives. Participants come into Phase I with a project that has some key assumptions already defined. In Phase I the project will be documented in the form of a project charter, which will include the business case for the effort. Phase I will also cover a powerful method (QFD) for transitioning the voice of the customer into design requirements.

Phase II : Methods for identifying alternatives and evaluating them

As far as evaluating design alternatives, the first pass is centered on the Pugh concept selection technique. The "total design" concept is introduced by Pugh (Pugh, 1991). According to the Pugh's definition, the total design is the necessary systematic activity, from the identification of the market/user need, to the selling of the successful product that satisfies the need-an activity that encompasses product, process, people and organization. A total design activity model is proposed such that it has a central core of activities all of, which are imperative for any design regardless of the domain. This core is called the design core and consists of market/user need, product design specification, conceptual design, detail design, manufacture and sales.

This will allow the DFSS team to select (or improve upon) the best alternatives. The next mode of evaluation is based on the FMEA process. Here teams will evaluate a selected design concept for potential failure modes so that they can be addressed early in the design effort.

PUGH concept selection method:

Following steps are critical to select an appropriate concept:

- Choose criteria (QFD)

- Create decision matrix (rows-criteria, columns-concepts)

- Select reference concept (can be one of the better concepts)

- Score (B=Better, W=Worse, S=Same)

- Repeat process with "winning" concepts (eliminating dogs)

- Weighted? (Analytical Hierarchy Process software)

Phase III: Design optimization techniques

The most powerful of these is Robust Design. This phase does more than identify best nominal settings for design parameters; it also provides a key set-up for the verification activities that come later.

Phase IV: Design optimization with Tolerance Design techniques

This technique provides a logical and objective basis for setting manufacturing tolerances. After the tolerance issues are addressed, the team will turn to the final verification and validation activities. These activities range from (1) ensuring that the product as designed meets the requirement set to (2) establishing process controls in manufacturing to ensure that critical characteristics are always produced correctly. Also in this phase the team creates a final cost/benefit report with key lessons learned. This enables long term improvement in the process.

DFSS is the way for companies to realize the full benefits of Six Sigma performance. DFSS has a substantial effect on long term profitability through improved products (and improved efficiencies) which results in increased customer satisfaction (and reputation) which results in improved market share which leads to increased profit potential.

To prepare people to support this DFSS process a four phase program has been developed, each phase is followed with expert consulting to ensure that the tools and methods are being properly applied.

An important ingredient for a successful DFSS project is clear role definition and solid role execution. The best process will not work if the players don't perform their parts well; this is especially true in a large scale, complex effort like DFSS.

3.3 Failure Mode Effect Analysis (FMEA)

The FMEA discipline was developed in the United States Military. Military Procedure MIL-P-1629, titled Procedures for Performing a Failure Mode, Effects and Criticality Analysis, is dated November 9, 1949. It was used as a reliability evaluation technique to determine the effect of system and equipment failures. Failures were classified according to their impact on mission success and personnel/equipment safety. The term "personnel/equipment", taken directly from an abstract of Military Standard MIL-STD-1629, is notable. The concept that personnel and equipment are interchangeable does not apply in the modern manufacturing context of producing consumer goods. The manufacturers of consumer products established a new set of priorities, including customer satisfaction and safety. As a result, the risk assessment tools of the FMEA became partially outdated. They have not been adequately updated since.

In 1988, the issued the ISO 9000 of business management standards. The requirements of ISO 9000 pushed organizations to develop formalized Quality Management Systems that ideally are focused on the needs, wants, and expectations of customers.QS9000 is the automotive analogy to ISO 9000. A Task Force representing Chrysler Corporation, Ford Motor Company, and General Motors Corporation developed QS 9000 in an effort to standardize supplier quality systems.

In accordance with QS 9000 standards, compliant automotive suppliers shall utilize Advanced Product Quality Planning (APQP), including design and process FMEAs, and develop a Control Plan.

Advanced Product Quality Planning standards provide a structured method of defining and establishing the steps necessary to assure that a product satisfies the customer's requirements. Control Plans aid in manufacturing quality products according to customer requirements in conjunction with QS 9000. An emphasis is placed on minimizing process and product variation. A Control Plan provides "a structured approach for the design, selection, and implementation of value-added control methods for the total system." QS 9000 compliant automotive suppliers must utilize Failure Mode and Effects Analysis (FMEA) in the Advanced Quality Planning process and in the development of their Control Plans.

FMEA is also a way to predict the consequences of failures in complex systems. FMEA has been a required part of many aerospace and military projects for many years. A number of support tools have been developed to aid in the time-consuming process of performing FMEA's. These tools generally collected data, kept it organized and produced reports according to Mil Stds. But the process of identifying the failure characteristics of individual components and determining the effects of those failures as they propagate across the systems in which they are embedded has been mainly a manual and expensive process. Powerful workstations and modern mathematical techniques are now making it possible to automate FMEA and make it a routine and affordable part of the engineering design process. This allows higher quality products to be produced faster, cheaper and more competitively. FMEA helps to identify potential catastrophic and critical failures of the critical real-time functions so that susceptibility to the failures and their effects can be reduced or eliminated from the system.

3.3.1 FMEA Process flow

Identify Functions
Identify Failure mode
Identify Effect of the Failure Mode
Determine Severity
Apply procedure for Potential Consequences
Identify possible causes
Determine Occurrence
Calculate Criticality
Identify Design or Process control
Determine Detection
PRN Calculation
Action to reduce

3.4 Risk Priority Number (RPN)

The RPN is a step that can be used to help prioritize failure modes for action. It is calculated for each failure mode by multiplying the numerical ratings of the severity, probability of occurrence and the probability of detection (effectiveness of detection controls). In general, the failure modes that have the greatest RPN receive priority for corrective action. The RPN should not firmly dictate priority as some failure modes may warrant immediate action although their RPN may not rank among the highest. In the example, the RPN would suggest that the light bulb would be of the highest priority, however, the realistic priority may be the cord because of the associated safety risks

It is a mathematical product of the numerical Severity, Occurrence, and Detection ratings.

RPN = $(S)* (O)* (D)$. This number is used to place priority on items than require additional quality planning.

Severity (redundancy)

Occurrence (change in design, or processes)

Detection (improve ability to identify the problem before it becomes critical)

4   CHAPTER FOUR

4.1 Six Sigma and leadership

One of the most important criteria for a successful implementation of Six Sigma is the executive management support. People, companies, industries, economies and nations will maintain leadership and a competitive edge only if they have a consistent mental attitude and a thirst for more effective ways to produce state-of-the-art products and services. The human element is finally taking advantage of an era of technological development to create a discipline based on the reality those higher levels of unbelievable quality at lower cost can be achieved to compete in a challenging global economy. Six Sigma is more than a road map for survival. It is the route to profitable growth. People create results. Involving all employees is essential to implementation of any successful program. Quality gurus have made it clear that the ultimate responsible for successful implementation of any quality program is the executive management. Juran believes that management is responsible for 85% of the problems in a company, as well as a primary factor in the achievements and success of a company. Based on this philosophy and the knowledge of business, management is responsible for total improvement in the company. Implementation of Six Sigma will not happen without great leadership, vision, planning and most of all, unity among all employees.

Based on the above definitions and considering the strong relationship between robust quality and management influence, certainly the rational leader will bring quality and consistency to the factory more than anybody else. When decisions from management are made haphazardly and randomly, productivity and quality suffer. Consistency of implementing a total quality culture and the zero defect goal is one of the most important tasks the leader needs to carry out. Consequently, it must be done in a consistent, focused and well planned manner. Managers can be assigned as required because they are abundant; however, there is not a job description in the personnel department for a leader. Leaders will create success and bring vision, goals, decision making ability and the often elusive stabilizing focus required. Leaders set the goals, build a quality organization, quality product and work relentlessly on reducing the failures throughout the required processes.

Quality leadership must create an environment that focuses on reducing variation in both processes and the organization as a whole. Robust quality requires adapting a more flexible, patient, and participate managerial style. This can require concentrated effort from anyone accustomed to ruling in an autocratic style. The results are well worth the effort. The leader must fully understand the philosophy and concept of product and process improvement on a continuing basis through teamwork and motivating people.

Motivation is the internal desire to act in order to meet felt needs. Leadership, while often dependent on a focal person who is motivated to lead, is centered on the leader providing climate for a group or groups within an organization. Quality heavily depends on the correct mix of the leader's style of management characteristics of the led and the situation. Classical definition usually questions how much power the organization gives the leader; however, the new approach is how much power the leader will obtain from correct decision making and from creating motivation in the organization.

A leader brings value to the organization and the organization will create value personnel, value decisions and ultimately, the successes and failure in the organization. Based on the above explanations what should we need to do in order to achieve this level of quality and leadership?

Developing a quality mentality requires a robust environment which needs to be created by the leadership, knowledge and planning. A leader brings motivation, measurement and employee involvement to the organization. There has been enough emphasis to quantify quality as a precision management tool so quality measurement, systems and organizational behavior can be developed. Leadership creates a quality attitude and quality culture which promotes a high level of creativity and motivation, but it is always associated with some risk. Without taking risks and without breaking the paradigms, the organization will not evolve, will not react to the market changes and will not develop itself. Good leadership can contribute substantially to high quality, while poor leadership is often a major cause of many quality problems in the organization.

4.2 Committed leadership

The Six Sigma approach is essentially about achieving business transformation and not surprisingly, it requires specific leadership qualities to make it happen. Committed leaders who have the clarity of purpose and drive to achieve breakthroughs in performance are a vital element in all Six Sigma programs.

The transformational leader in the Six Sigma approach is often described as having an edge. Frequently cited examples include Jack Welch (GE), Larry Bossidy (Allied Signal) and Bob Galvin (Motorola). The commitment to Six Sigma is evident through their behavior, drive, energy and commitment to provide the necessary resources. It would be unwise to assume that this rather 'hard' model of leadership represents some unified theory Leadership research is littered with theories - trait theories, behavioral theories, situational/contingency theories etc. Transformational leaders must certainly have that edge to make tough strategic decisions but other qualities, including a genuine concern for the people being 'transformed', must play a key role in delivering sustained success.

4.3 Change Acceleration Process (CAP)

The CAP or Change Acceleration Program originated at GE at the direct request of CEO Jack Welch, who commissioned a design team directed by D. Ulrich, Noel Tichey and Steve Kerr to identify the best knowledge and practices available on leading change and then to create the approach to be used by all GE Managers. Just the cost of researching and designing this Change Acceleration

Program at GE was over $5 mil. CAP was designed also to help each initiative to become a permanent and vital part of the GE corporate culture.

This powerful tool began the GE transformation process by engaging employee teams in quickly dealing with problems in their own work areas, streamlining processes and getting rid of unvalued work-Workout! With the many small, fast moving changes this generated across the company. GE needed a more strategic, comprehensive approach to planning and executing change that could bring together smaller, disconnected efforts in a coordinated, more focused way and make their execution faster and smoother. The CAP program very successfully fulfilled that need and continued to grow and, years later, incorporated the Six Sigma effort. It was also used extensively with suppliers, customers and other GE partners, and even "bartered" for concessions and contractual agreements with those partners.

The model and tools are very hands-on and application oriented, rather than philosophical or academic. They provide step by step, practical "recipes" for Focusing and Executing a change initiative in a way that greatly increases implementation speed by systematically building buy-in at every level of the organization.

This simple approach is built on the assumption that successful change execution requires two streams of activity:

1. The Quality of the change-what we plan to do-we call the "Q", and,

2. The organizational, cultural, political and people aspects of the change-the commitment or the "A" which is the Acceptance of the change. Then the effectiveness (E) of the change will be defined as:

$$E = Q \times A$$

Most organizations quickly recognize their effectiveness at determining the "Q", while often putting off, stumbling or even ignoring building buy-in and commitment-the "A", leading to the ultimate failure of most of their change efforts.

By examining the reasons for these past change failures and successes, participants realize that almost all the variables that determine the success of their past change initiatives relate to the organization, cultural and political aspects of their changes. The quality of the "Q" is seldom the cause of failure.

The Change Acceleration Process (CAP) can be either a constructive team effort where employees embrace the new concepts and processes or it can be considered a "phase" of management planning and therefore ignored or fought every step of the way.

The "buy-in" to the implemented changes is therefore built into the CAP process. Employees are encouraged and taught how to research and identify Best Practices and also how to perform Industry Benchmarking Studies. The most important factor is that the organizations need for the change must exceed than the resistance that the people and the infrastructure of the organization has. We are asking people to change their behavior and do something, such as to buy and pay for goods and services differently than they have done so in the past. A basic rule of change management is to "communicate, communicate, and communicate". You can (almost) never share enough information with any of the stakeholders you identify.

The key steps through the successful CAP implementation are:

1. Develop your communications plan

2. Choose your audience or stakeholders

3. Develop the message for each audience or stakeholder group

4. Determine the best time for the message to be delivered

5. Utilize the best methods for message delivery

6. Communicate regularly

7. Use clear and concise messages

8. Keep the message simple

9. Explain the benefits – what's in it for me?

10. Use positive undertones

11. Answer questions in the first paragraphs

12. Use Senior Exec, and manager quotes

13. Involve your suppliers

4.3.1 Developing communication plan

Before you develop your communications plan, think about your company culture. How is the plan viewed by the various stakeholders? Is the program viewed positively? Or is it regarded with hesitation and skepticism? To encourage card acceptance, some organizations use only a "pull" marketing strategy for their program. That is to say, they rely on success stories, word of mouth, and the office grapevine to encourage card usage and spur continued growth. In these organizations, internal marketing consists of videos, newsletters, give-aways, and displays. The excitement of using a P-Card or travel card is what "pulls" other employees with the need to buy into the program.

4.4 Stakeholder

What information & When to communicate?

Early on in your program you should identify stakeholders, or target audiences with a particular interest in the success of the program. Be certain to tailor communications to these groups so that their questions are answered and the "What's in it for me" is addressed. The important factor is that you have a communication plan that is appropriate for each stakeholder group and that you follow through on the action items identified. Program administrators have, however, offered the suggestions noted below regarding program communications in general.

Regardless of how long your program may have been in place, continue to ask customers and involved employs and suppliers for their ideas on how to make program enhancements and improve communications with them.

Along those lines, communicating about and introducing the change ideas to employees should begin as early as employee orientation, as not before. Just like other new activities, information sharing should be incorporated into the orientation process. GE Corporate Payment Services utilizes a tool called the Partnership Business Plan (PBP) to formally assess program progress against mutually set performance goals, obtain customer feedback, and benchmark your program against others of similar size. Quality measures, supplier strategies, and discussion of your overall communications plan are also included in the Partnership Business Plan. Your involvement in this process is highly recommended as the PBP is a highly effective communications and program management tool. The communications plan extends beyond your own organization. Attending user conferences, industry seminars, monthly user teleconferences, workshops, and professional association meetings are all opportunities to discuss your program and / or formally present an overview of what you've done and how you've done it. It is an excellent idea to have a public or externally focused card program presentation prepared (and regularly updated) so that as you are asked to do so, you can teach others about lessons learned and goals achieved. Doing so benefits everyone, including you. Keep in mind that to share information about the program with all stakeholders on a consistent, regular basis. Tailor your messages depending on the audience, but send selected communications tips and plans to every stakeholder group. Be sure to regularly survey users and others involved in the program for their feedback and suggestions. Some of the best ideas for card usage, program communications, and policy revisions come from the stakeholder themselves. On an external note, "shamelessly steal and willingly share" ideas about your program, and then remember to "communicate, communicate, communicate" with your colleagues and co-workers. Figure…. Illustrates the duration of the change process from the beginning to the end. Basically with any types of changes there will be a "Valley of despair" which lots of organization do not realize and fail to complete the full implementation of the process. Bridges, Enhancements from Val Larson 2002.

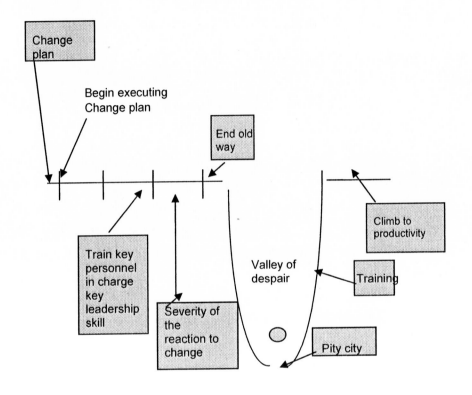

5 CHAPTER FIVE

5.1 *Essential Tools for Six Sigma implementation*

5.1.1 Affinity Diagram:

The purpose of affinity diagram is to organize large groups of information into meaningful categories.

The affinity diagram helps break old patterns of thought, reveal new patterns, and generates more creative ways of thinking. The affinity diagram helps organize the team's thoughts most effectively when: the issues seem too large and complex; you need to break out of old, traditional ways of thinking; everything seems chaotic; or there are many customer's requirements. The affinity diagram and relationship diagram offer interesting tools for gathering and organizing the information gathered from your customer. These can be used to gather, correlate, and relate huge quantities of information (especially written or verbal comments). The affinity diagram helps to naturally group ideas or your customer's valid requirements and show the relationships between items and groups. The affinity diagram helps you gather and group large amounts of language (e.g., needs, wants, wishes, ideas, and opinions) into natural relationships. This more organic and creative approach to understanding the user's needs is also a useful tool for object-oriented analysis. The affinity diagram helps organize the team's thoughts most effectively when: the issues seem too large and complex to grasp, not simple or immediate you need some way to break out of old, traditional ways of thinking facts or thoughts are chaotic you need to quickly uncover your customer's requirements. The process of affinity diagram is to state the issue to be examined in broad terms: What are the issues surrounding or involving. the delivery of very low defect products or services the delivery of very low defect products or services reducing cycle time reducing waste or rework. Figure 4 show a simple affinity diagram.

94

Affinity Diagram

"Figure 5"

Generate and record ideas using Post-it notes. Begin sticking them on a wall or large sheet of easel papers where everyone can see them. Ensure that everyone is included.

Ask for a headline to describe each thought. Note the contributor's initials.

Arrange the cards in related groupings. As you generate ideas, the person at the board may begin grouping the available notes as they are offered and keep the intensity of note generation going as long as possible.

Complete the groupings. Involve the group in clustering the notes into 6-10 related groupings. Have everyone stand and do this silently. Be prepared for some loners. Avoid forcing them into a group. Some notes may need to be duplicated for different groupings.

Choose a word or phrase that captures the intent of each group and place it at the top as a header card. If there isn't one already, then create one with a word or phrase that does capture the intent.

5.1.1.1 *How to Conduct an Affinity Sort*

Conduct a brainstorming session on the topic under investigation.

- Clarify the list of ideas. Record them on small cards or Post-It notes.

- Randomly lay out cards on a table, flipchart, wall, etc.

- Without speaking, sort the cards into "similar" groups based on your gut reaction.

- If you don't like the placement of a particular card, move it. Continue until consensus is reached.

- Create header cards consisting of a concise 3-5 word description; the unifying concept for the group. Place header card at top of each group.

- Discuss the groupings and try to understand how the groups relate to each other

5.1.2 Relationship Diagram:

The purpose of relationship diagram is to identify cause-effect or method-objective relationships among related items in a system when the interactions are complex or involved.

The relationship diagram, in contrast to the affinity diagram which only shows logical groupings, helps map the logical relationships between the related items uncovered in the affinity diagram. The relationship diagram shows cause and effect relationships among many key elements. It can be used to identify the causes of problems or to work backward from a desired outcome to identify all of the causal factors that would need to exist to ensure the achievement of an outcome. Figure 6 shows a simple relationship diagram.

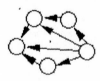

"Figure 6"

The key factors affecting the main drivers can then be examined using the tree diagram.

Process: State the problem or issue under discussion—software defects, customer retention, process steps, whatever. Write this on a Post-it. the key issue at the center of a white board or on one side or the other.

Generate cause-effect issues by brainstorming or by using the notes generated for the affinity diagram (except the header notes) and arrange them according to cause-effect.

(Note: There should be more than nine and less than fifty notes when completed, otherwise the problem is either too simple or too complex for this method). Draw one-way arrows to indicate the cause-effect relationship among all of the components of the diagram. Avoid two way arrows; decide which component has the most influence and draw the arrow in one direction only. Identify the key issues (ones with the most arrows coming out of them) with darker lines or other shading.

5.1.3 Tree Diagram:

The purpose of tree diagram is to systematically link ideas, targets, objectives, goals, or activities in greater and greater detail. The tree diagram can map specific tasks to primary and secondary goals. It maps the methods required to achieve corporate goals. The tree diagram shows the key goals, their sub-goals, and key tasks. It can help identify the sequence of tasks or functions required to accomplish an objective. The tree diagram can help translate customer desires into product characteristics. It can also be used like an Ishikawa diagram to uncover the causes of a particular problem. Figure 7 shows a simple tree diagram.

Tree Diagram

"Figure 7"

The process of tree diagram is to develop a clear statement of the problem, issue, or objective to be addressed.

Place it on the left side of a board, wall, or easel and work toward the right. Brainstorm all of the sub-goals, tasks, or criteria necessary to accomplish or resolve the issue.

Repeat this process using each of the sub-goals until only actionable tasks or elements remain. Check the logic of the diagram in the same way as the Ishikawa: Start at the right and work your way back to the left by asking: If we do this, will it lead to the accomplishment

5.1.4 Matrix Diagram

The purpose of matrix diagram is to compare two or more groups of ideas, determine relationships among the elements, and make decisions. The matrix diagram helps prioritize tasks or issues in ways that aid decision making; identify the connecting points between large groups of characteristics, functions, and tasks; or show the ranking or priority of in an interaction.

Combined with tree diagrams, prioritization matrices can rank various choices in terms of impact on the customer, reduction in cycle time, defects, costs, and so on.

Matrices can be used in many ways to show relationships. They can be shaped like an L, a T, an X, or a three-dimensional, inverted Y. The L-shaped matrix helps display relationships among any two different groups of people, processes, materials, machines, or environmental factors. The T-shaped diagram is simply two L-shaped diagrams connected together showing the relationships of two different factors to a common third one. The Y-shaped matrix helps identify interactions among three different factors. The X-shaped matrix (two T's back to back) is occasionally useful. Figure 8 shows a simple matrix diagram.

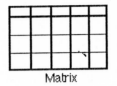

Matrix

"Figure 8"

The process of matrix diagram is to generate two or more sets of characteristics to be compared. Use tree diagrams or brainstorming. Choose the proper matrix to represent the interactions (L, T, X, Y). Put the characteristics on the axes of the matrix. Rank the interactions from 1 (low) to 5 (high)

5.1.5 Arrow Diagram

The purpose of arrow diagram is to show the paths to complete a project, find the shortest time possible for the project, and graphically display simultaneous activities.

The arrow diagram is closely related to a CPM (critical path method) or PERT (program evaluation and review technique) diagram. It is also known as an activity network diagram. It can be used to plan the schedule for any series of tasks and to control their completion. The arrow diagram removes most of the complexity of CPM and PERT methods, and retains the flow from task to task and the timing required for each task. It does, however, require that you know what each task is and how long it takes. Without such knowledge, it's difficult to develop the arrow diagram. Figure 9 shows a simple arrow diagram.

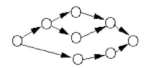

"Figure 9"

The process of arrow diagram is to brainstorm all of the tasks required to complete a given project, including the estimated time required for each task. Again, note cards are useful for this process.

Sequence all of the cards from the start to finish, removing duplications, and adding new ones as additional tasks are identified and placing parallel activities where they belong. Reevaluate the shortest, longest, and average estimated times for each task and identify the longest path through the diagram. Use the diagram to track progress of each activity throughout the project life cycle. Any time an element gets in jeopardy, it may need to be examined and necessary resources shifted to complete it.

5.1.6 Process Mapping

As it was mentioned in chapter 2, the process map is the visualization of the entire process. Everything a company does involves processes and these have to be identified, analyzed and improved upon for the ultimate corporate good. The map shows where the strengths and weaknesses are and leads to the reduction of defects.

5.1.7 Balanced Scorecard

Most organizations after having articulated and identified their various objectives are unable to communicate the strategies of the organization. One of the most effective methods for communicating the strategy of the organization is building a strategy map encompassing the now widely adopted Kaplan and Norton's Balanced score card spanning the four perspectives. Answering the questions related to the perspectives helps understand the strategy better and also build a good strategy map. The strategy map helps provide the vital cause and effect linkages in an organization and helps link the Business processes to the strategic destination of the organization. Before attempting to build the map it is essential for the organization to identify all its core processes and support processes as they help in completing the strategy map.

The innovative ideas and principles of the Balanced Scorecard were first described in the book "The Balanced Scorecard - Translating Strategy into Action" by Prof. Robert S. Kaplan from Harvard Business School and Dr. David P. Norton from Renaissance Solutions Inc. back in 1996.

It was the result of a group effort by representatives from a dozen corporations to come up with a better performance measurement system relying less on mere financial measures.

The basic idea is to focus the organization on metrics that matter as seen from a strategic point of view. To avoid focusing only on short term financial measures the scorecard comprises metrics from areas such as customer, internal processes, and learning and growth perspectives respectively. The process of "translating strategy into action" involves turning the company's strategic vision into clear and understandable objectives within all of those perspectives mentioned above.

In the financial perspective of the Balanced Scorecard, focus is on how the company should position itself to be considered an attractive and exciting investment to its owners. For a profit maximizing organization this is the prime reason for being, at least in theory.

To achieve these financial objectives the company will have to bring some kind of value to its customers (the value proposition) that the customer considers to be higher than the price paid.

Implementing a Balanced Scorecard solution involves a number of steps:

a. Build an understanding of the prime drivers of performance in your organization.

b. Develop a consensus on corporate direction and principal strategic objectives.

c. Translate your knowledge of drivers and your strategic objectives into a balanced set of top level measures (often called Key performance

Indicators or KPI's) that can be used to gauge both current and future performance.

d. Develop a system for monitoring, analyzing and reporting your key performance measures. The measures chosen normally include a mixture of indicators relating to financial performance, customer satisfaction and market share, internal processes and finally to innovation in your organization.

5.1.7.1 *Making decision based on Data*

Managers and coaches can use the information that is generated in the initial balanced scorecard sessions to really examine "why" they are doing what they are doing-to reassess and determine the critical few business results that form the top line of a balanced scorecard. This focus helps management at every level tackle one of the most difficult problems a company faces... establishing priorities among multiple competing, sometimes confusion, and often seemingly equally valuable results targets. The first benefit is the move toward clarity in the weighting exercises individual managers and leaders go through in assessing the various objectives against the return on investment and near-term, longer-term impact by aligning company-wide focus in a systematic way.

Once the targets are defined and broken into 4-6 key business objectives at the executive levels, those key targets (e.g., lets say for purposes of this example: Achieve a 12% increase in profitability over the Year 2003 through weighted emphasis on 1) Production, 2) Sales, 3) Service, 4) Knowledge (human capital) Retention). The next levels of the organization, Divisions, Departments, Units, Work Groups or Teams then look at their particular objectives through a cascading rollout against this core list and assess what they need to do to achieve the critical few results that they have control over that helps achieve the overall results. A Team such as Sales/Marketing may link back to the company scorecard through their efforts in 1) sales and 2) retention, for example, but not necessarily in Production, and Service.

The Sale/Marketing Team then defines 5 key goals/results targets for Sales and 2 result targets for Retention. Just as the company came up with relative weights for their 5 core areas, the sales/marketing department might weigh sales at 60% and retention as 40%.

Once such an exercise is done, then the next level defines objectives that are assigned to the key areas-the two core targets of Sales and Retention form the basis a series of objectives for the team. Individual members of the sales team then assess what their role is in achieving desired outcomes through the PM behavioral link-the Performance Process Improvement Plans (PIPs) that they will implement to help achieve the objectives of the unit. They can see clearly how they are aligned all the way to corporate profitability and it takes the "guess work" out of what are the critical few PIP behaviors for individual performers. A department may have one or two other key needs such as process improvement or performance systems redesign to make sure they can do what is required, but even those can be linked to increased sales effectiveness, in keeping with this example, and directly linked to individual performers activities to help with process improvement.

In many companies without such an alignment, managers determine, often with the employee, that the individual should do 6 things to improve performance. That may be true, but such coaching/managing pinpointing sessions can be increased in effectiveness 10 fold by assuring that what Tom, Joe, and Mary need to do is derived from and directly impact the core objectives of the team and, in turn, of real benefit to the company.

The behaviors that an individual does can be designed in such a way that both his/her developmental needs and the company's core objectives are achieved. All too often at the individual performer level, behavioral pinpoints are done to ensure individual effectiveness, not group, team, company-wide synergy.

When the concept of Team Results is discussed, it is common to assume that those "group results" will be achieved somehow by the act of being part of a team. The balanced scorecard and individual performance objectives much better ensures that the team will achieve its results. Following is the illustration of Balance Scorecard for the four perspectives, financial, customer, internal and innovation and learning.

"Figure 10, illustration of Balance Scorecard".

For example, let's say that the example above is now down to a Sales Team, all of whom are expected to sell 40% more of Product X within the next 6 months. That objective, tied to 4 other Sales Team objectives will allow the team to contribute their part of the 60% the Sales/Marketing Dept. Has committed to the company score card in the area of sales (another 40% of the Company-wide Sales Objective will come from New Product Development, another team located in the Research and Development Division of the Company). In the meantime, the Sales/ Marketing Department then takes the target of 40% selling objective for Product x and assigns to each Sales Rep target sales objectives according to relative skills, knowledge of the product, geographic territory, and so on. Individual objectives are then designed around behavioral targets, PIPs, to help achieve the Team objective. Once the objectives have cascaded to this level, there can be development focus and objectives that may get to how the individual interacts with others on the team to achieve sales (improving interpersonal relations or personal style issues of timeliness or other factors that may impact his/her sales success). This process is really focused on individual alignment for success.

A core concept of PM effectiveness is that neither Team PIPs nor individual PIPs should be determined independent of the key drivers of the company. Just as with behavioral pinpointing when used at the individual level to help the individual sort through multiple targets for improvement or maintenance, this balanced scorecard process encourages managers to look at selecting the critical few results pinpoints from the relative gain or "new effort" or "importance against the competitive landscape"...or other key business drivers a company faces and how the Team and its members contributes to those key areas. This process allows both managers and coaches to then pinpoint how individuals can really help drive success and increase the likelihood that all employees will feel valued and a key part of the company's accomplishments.

Such an approach lessens the distractions of competing priorities and so many of the mistakes of current managers and coaches in addressing the individual performer through traditional performance appraisal that is all too often far removed from the key actions/behaviors required to achieve company results. If done correctly, everyone from the receptionist or cleaning personnel to the CEO can see how his/her performance leads to overall organizational success.

5.1.7.2 *Benefits of Balance Scorecard*

a. Valid and meaningful data related to key measures

b. Vehicle for improving communication channels with customers and obtaining valuable feedback

c. Process for organizing and analyzing performance data so that strengths and weaknesses can be identified

d. Tool to increase visibility of goals in the eyes of senior management to focus attention on successes and areas needing improvement

e. Establish performance trends, and tract progress

f. Flexible process for self-assessment with minimal reporting requirements

g. Develop actual information-gathering tools.

5.1.8 Supplier Input Process Output Customer (SIPOC)

SIPOC stands for suppliers, inputs, process, output, and customers. You obtain inputs from suppliers, add value through your process, and provide an output that meets or exceeds your customer's requirements.

Supplier-Input-Process-Output-Customer: Method that helps you not to forget something when mapping processes.

A SIPOC diagram is a tool used by a team to identify all relevant elements of a process improvement project before work begins. It helps define a complex project that may not be well scoped, and is typically employed at the Measure phase of the Six Sigma DAMIC methodology. It is similar and related to process mapping and In/Out of scope' tools, but provides additional detail.

The tool name prompts the team to consider the Suppliers (the 'S' in SIPOC) of your process, the Inputs (the 'I') to the process, the Process (the 'P') your team is improving, the Outputs (the 'O') of the process, and the Customers (the 'C') that receive the process outputs. In some cases, Requirements of the Customers can be appended to the end of the SIPOC for further detail. The SIPOC tool is particularly useful when it is not clear: Who supplies Inputs to the process? What specifications are placed on the Inputs? Who are the true Customers of the process? What are the Requirements of the customers?

A SIPOC Model is built for each business process within the scope of the Six Sigma project. The SIPOC Model establishes the boundaries of a particular business process. SIPOC is an acronym for Suppliers, Inputs, Processes, Outputs and Customers and shows how these business entities interact at the boundary of the process. A business process is a time bounded set of activities that consumes resources and produces products and/or services. Each process's SIPOC model provides the reader with a firm idea of the beginning of the process (by showing its resources and suppliers), and the ending points of the process (by showing its outputs and customers). The SIPOC Models provide a process-driven approach to divide the entire scope of the Six Sigma project into manageable partitions.

The SIPOC Models define the high-level process participants that are included in the scope of this Six Sigma project. The SIPOC Model is created after a Business Interaction Model, or in place of, if the scope of analysis is confined to a single business process. A SIPOC model can be built using the specific process under study as its domain.

5.1.8.1 *A SIPOC and business process*

The business processes identified in the SIPOC Models are associated with the critical business goals. This serves as a cross-checking mechanism to ensure a complete set of goals and processes have been identified. As with the deliverables, business processes are associated to goals using an association matrix. Typically, the business's core processes (those, which produce the products and services, delivered to customers), support customer-driven goals. Support processes (those supporting the core processes, such as Manage Finances) may address only a few customer-driven goals.

The SIPOC technique can be used to identify all relevant elements of a process improvement project before work begins. It helps define a complex project that may not be well scoped, and is typically employed at the Measure phase of the Six Sigma DMAIC methodology. It is related to Process Mapping and "In/Out Of Scope" tools, but provides additional important detail.

Tools can be used for capturing SIPOC information. Typically, the tool prompts the user for the name of the process step or activity (the "P"), for the inputs required (the "I"), the outputs created (the "O"), the source or supplier of the input (the "S"), and finally the customer or destination for the outputs (the "C").

In some cases, requirements can be appended to the end of the SIPOC for further detail.

5.1.9 Quality Function Deployment or House of Quality (QFD)

QFD was developed in Japan in the late 1960s by Professors Shigeru Mizuno and Yoji Akao. At the time, statistical quality control, which was introduced after World War II, had taken roots in the Japanese manufacturing industry, and the quality activities were being integrated with the teachings of such notable scholars as Dr. Juran, Dr. Kaoru Ishikawa, and Dr. Feigenbaum that emphasized the importance of making quality control a part of business management, which eventually became known as TQC and TQM.

The purpose of Professors Mizuno and Akao was to develop a quality assurance method that would design customer satisfaction into a product *before* it was manufacture red. Prior quality control methods were primarily aimed at fixing a problem during or after manufacturing.

It is a powerful tool for determining and prioritizing the Critical To Quality (CTQs). Metrics which directly influence Voice of Customers (VOC). QFD or house of quality has been developed through a brainstorming and assignment of priority numbers and importance weight to each metric. Each metric carries a weight in correlation with each CTQ. QFD uses a series of matrices to document information collected and developed and represent the team's plan for a product. The QFD methodology is based on a systems engineering approach consisting of the following general steps:

5.1.9.1 *Collecting Customers' Data*

1. How to evaluate and analyze customer survey and customer data and feedback?

Consider customer requirement documents, requests for proposals, requests for quotations, contracts, customer specification documents, customer meetings/interviews, focus groups/clinics, user groups, surveys, observation, suggestions, and feedback from the field. Consider both current customers as well as potential customers. Pay particular attention to lead customers as they are a better indicator of future needs. Plan who will perform the data collection activities and when these activities can take place. Schedule activities such as meetings, focus groups, surveys, etc.

2. Identify required information. Prepare agendas, list of questions, survey forms, focus group/user meeting presentations.

3. Determine customer needs or requirements using the mechanisms described in step 1. Document these needs. During customer meetings or focus groups, ask "why" to understand needs and determine root needs. Consider spoken needs and unspoken needs. Extract statements of needs from documents. Summarize surveys and other data. Use techniques such as ranking, rating, paired comparisons, or conjoint analysis to determine importance of customer needs. Gather customer needs from other sources such as customer requirement documents, requests for proposals, requests for quotations, contracts, customer specification documents, customer meetings/interviews, focus groups, product clinics, surveys, observation, suggestions, and feedback from the field.

4. Use affinity diagrams to organize customer needs. Consolidate similar needs and restate. Organize needs into categories. Breakdown general customer needs into more specific needs by probing what is needed.

Maintain dictionary of original meanings to avoid misinterpretation. Use function analysis to identify key unspoken, but expected needs.

5. Once needs are summarized, consider whether to get further customer feedback on priorities. Undertake meetings, surveys, focus groups, etc. to get customer priorities. State customer priorities using a 1 to 10 rating. Use ranking techniques and paired comparisons to develop priorities.

6. Organize customer needs in the Product Planning Matrix. Group under logical categories as determined with affinity diagramming.

7. Establish critical internal customer needs or management control requirements; industry, national or international standards; and regulatory requirements. If standards or regulatory requirements are commonly understood, they should not be included in order to minimize the information that needs to be addressed.

8. State customer priorities. Use a 1 to 10 rating. Critical internal customer needs or management control requirements; industry, national or international standards; and regulatory requirements, if important enough to include, are normally given a rating of 7.

9. Develop competitive evaluation of current company products and competitive products. Use surveys, customer meetings or focus groups/clinics to obtain feedback. Rate the company's and the competitor's products on a 1 to 10 scale with "10" indicating that the product fully satisfies the customer's needs. Include competitor's customer input to get a balanced perspective.

10. Review the competitive evaluation strengths and weaknesses relative to the customer priorities. Determine the improvement goals and the general strategy for responding to each customer need. The Improvement Factor is "1" if there are no planned improvements to the competitive evaluation level. Add a factor of .1 for every planned step of improvement in the

competitive rating, (e.g., a planned improvement of going from a rating of "2" to "4" would result in an improvement factor of "1.2". Identify warranty, service, or reliability problems & customer complaints to help identify areas of improvement.

11. Identify the sales points that Marketing will emphasize in its message about the product. There should be no more than three major or primary sales points or two major sales points and two minor or secondary sales points in order to keep the Marketing message focused. Major sales points are assigned a weighting factor of 1.3 and minor sales points are assigned a weighting factor of 1.1.

12. The process of setting improvement goals and sales points implicitly develops a product strategy. Formally describe that strategy in a narrative form. What is to be emphasized with the new product? What are its competitive strengths? What will distinguish it in the marketplace? How will it be positioned relative to other products? In other words, describe the value proposition behind this product. The key is to focus development resources on those areas that will provide the greatest value to the customer. This strategy brief is typically one page and is used to gain initial focus within the team as well as communicate and gain concurrence from management.

13. Establish product requirements or technical characteristics to respond to customer needs and organize into logical categories. Categories may be related to functional aspects of the products or may be grouped by the likely subsystems to primarily address that characteristic. Characteristics should be meaningful (actionable by Engineering), measurable, practical (can be determined without extensive data collection or testing) and global. By being global, characteristics should be stated in a way to avoid implying a particular technical solution so as not to constrain designers. This will allow a wide range of alternatives to be considered in an effort to

better meet customer needs. Identify the direction of the objective for each characteristic (target value or range, maximize or minimize).

14. Develop relationships between customer needs and product requirements or technical characteristics. These relationships define the degree to which as product requirement or technical characteristic satisfies the customer need. It does NOT show a potential negative impact on meeting a customer need - this will be addressed later in the interaction matrix. Consider the goal associated with the characteristic in determining whether the characteristic satisfies the customer need. Use weights (for example using 5-3-1 weighting factors) to indicate the strength of the relationship - strong, medium and weak. Be sparing with the strong relationships to discriminate the really strong relationships.

15. Perform a technical evaluation of current products and competitive products. Sources of information include: competitor websites, industry publications, customer interviews, published specifications, catalogs and brochures, trade shows, purchasing and benchmarking competitor's products, patent information, articles and technical papers, published benchmarks, third-party service & support organizations, and former employees. Perform this evaluation based on the defined product requirements or technical characteristics. Obtain other relevant data such as warranty or service repair occurrences and costs.

16. Develop preliminary target values for product requirements or technical characteristics. Consider data gathered during the technical evaluation in setting target values. Do not get too aggressive with target values in areas that are not determined to be the primary area of focus with this development effort.

17. Determine potential positive and negative interactions between product requirements or technical characteristics using symbols for strong or medium, positive or negative relationships. Too many positive interactions

suggest potential redundancy in product requirements or technical characteristics. Focus on negative interactions - consider product concepts or technology to overcome this potential trade-offs or consider the trade-off in establishing target values.

18. Calculate importance ratings. Multiply the customer priority rating by the improvement factor, the sales point factor and the weighting factor associated with the relationship in each box of the matrix and add the resulting products in each column.

19. Identify a difficulty rating (1 to 5 point scale, five being very difficult and risky) for each product requirement or technical characteristic. Consider technology maturity, personnel technical qualifications, resource availability, technical risk, manufacturing capability, supply chain capability, and schedule. Develop a composite rating or breakdown into individual assessments by category.

20. Analyze the matrix and finalize the product plan. Determine required actions and areas of focus.

21. Finalize target values. Consider the product strategy objectives, importance of the various technical characteristics, the trade-offs that need to be made based on the interaction matrix, the technical difficulty ratings, and technology solutions and maturity.

22. Maintain the matrix as customer needs or conditions change. Following is the illustration of HoQ or QFD structure.

"Figure 11, standard QFD structure"

114

5.1.10   Alternative approach

Consider not only the current approach and technology, but other alternative concept approaches and technology. Use brainstorming. Conduct literature, technology, and patent searches. Use product benchmarking to identify different product concepts. Develop derivative ideas. Perform sufficient definition and development of each concept to evaluate against the decision criteria determined in the next step.

Evaluate the concept alternatives using the Concept Selection Matrix. List product requirements or technical characteristics from the Product Planning Matrix down the left side of the Concept Selection Matrix. Also add other requirements or decision criteria such as key unstated but expected customer needs or requirements, manufacturability requirements, environmental requirements, standards and regulatory requirements, maintainability / serviceability requirements, support requirements, testability requirements, test schedule and resources, technical risk, business risk, supply chain capability, development resources, development budget, and development schedule.

Carry forward the target values for the product requirements or technical characteristics from the Product Planning Matrix. Add target values as appropriate for the other evaluation criteria added in the previous step. Also bring forward the importance ratings and difficulty ratings associated with each product requirement or technical characteristic from the Product Planning Matrix. Normalize the importance rating by dividing the largest value by a factor that will yield "5" and post this value to the "Priority" column. Review these priorities and consider any changes appropriate since these are the weighting factors for the decision criteria. Determine the priorities for the additional evaluation criteria added in the prior step. List concepts across the top of the matrix.

Perform engineering analysis and trade studies. Rate each concept alternative against the criteria using a "1" to "5" scale with "5" being the highest rating for satisfying the criteria. For each rating, multiply the rating by the "Priority" value in that row. Summarize these values in each column in the bottom row. The preferred concept alternative(s) will be the one(s) with the highest total.

For the preferred concept alternative(s), work to improve the concept by synthesizing a new concept that overcomes its weaknesses. Focus attention on the criteria with the lowest ratings for that concept ("1's" and "2's"). What changes can be made to the design or formulation of the preferred concept(s) to improve these low ratings with the product concept? Compare the preferred concept(s) to the other concepts that have higher ratings for that particular requirement. Are there ways to modify the preferred concept to incorporate the advantage of another concept?

6 CHAPTER SIX

6.1 *Costs, Limitations and Challenges.*

One of the most important parts of Six Sigma implementation is the executive management commitment to the program. Since Six Sigma has a direct impact on the companies' bottom line financial numbers, it is a MUST that the management provides their full support.

No technology is right for every situation, and each technology has associated costs. This section points out these limitations and costs. Some examples of the kinds of costs and limitations that a technology may possess are the following: a technology may impose an otherwise unnecessary interface standard; it might require investment in other technologies it might require investment of time or money; or it may directly conflict with security or real-time requirements. Specific items of discussion include what is needed to adopt this technology (this could mean training requirements, skill levels needed, programming languages, or specific architectures) how long it takes to incorporate or implement this technology barriers to the use of this technology reasons why this technology would not be used Institutionalizing Six Sigma into the fabric of a corporate culture can require significant investment in training and infrastructure. There are typically three different levels of expertise cited by companies: Green Belt, Black Belt Practitioner, and Master Black Belt. Each level has increasingly greater mastery of the skill set. Roles and responsibilities also grow from each level to the next, with Black Belt Practitioners often in team/project leadership roles and Master Black Belts often in mentoring/teaching roles. The infrastructure needed to support the Six Sigma environment varies. Some companies organize their trained Green/Black Belts into a central support organization. Others deploy Green/Black Belts into organizations based on project needs and rely on communities of practice to maintain cohesion.

Based on the number of articles written this past year about GE and its CEO, Jack Welch, GE has now become the standard bearer for how Six Sigma is implemented to successfully drive positive bottom line impact along with recognized "World Class" status. Other highly respected and successful companies such as SONY are benchmarking off of GE and implementing a similar strategy.

The companies mentioned thus far are certainly well known for their engineering and manufacturing excellence. What is not as well known is their view of the importance of Six Sigma in non-manufacturing or transactional areas. Bob Galvin, former President and CEO of Motorola, has stated that the lack of initial Six Sigma emphasis in the non-manufacturing areas was a mistake that cost Motorola at least $5 Billion over a 4-year period. It is common these days to hear comments like, "Yes, Company X has a great product, but they sure are a pain to do business with!" Consequently, Jack Welch is mandating Six Sigma in all aspects of his business, most recently in sales and other transactional (non-manufacturing) processes. Unfortunately, the typical response from non-manufacturing employees has been, "We're different. Six Sigma makes sense for manufacturing but does not apply to us!" This is simply an excuse in order to avoid being held to the same accountability standards as manufacturing.

6.2 The Six Sigma challenge

Once an organization decides to implement a Six Sigma program, it must impart the challenge to every employee. This includes not only people close to production — where indexes and measurements are relatively easy to implement on physical processes — but also administrative and service providers.

Through an executive directive, the organization establishes its Six Sigma challenge, vision, customer satisfaction promise, goal and new measurement indexes. The directive distinguishes between former business policies and the new challenge of working toward excellence. It establishes a common goal for all employees in the organization: reduce variability (i.e., standard deviation) in everything they do. The directive requires all employees to participate in a day-long course outlining the "Five W's" of Six Sigma. This course explains the who, what, where, why and when of the organization's new way of doing business.

In a Six Sigma organization, employees assess their job functions with respect to how they improve the organization. They define their goals, or the ideal of excellence in their roles, and quantify where they are currently — their status quo — with respect to these ideals. Then they work to minimize the gap and achieve Six Sigma by a certain date.

Individuals in the finance department, mail room, human resources, purchasing and everywhere else also are challenged to achieve Six Sigma in everything they do, cumulatively bringing excellence to the organization as they achieve individual excellence in their jobs.

For an organization to reach Six Sigma successfully, the program must define a standard approach. If the approach is left undefined, too many individuals will spend too much time engineering and reengineering it. Standardizing a methodology to achieve Six Sigma allows individuals to focus on reducing the standard deviation within their individual projects rather than obsessing over method. It also establishes a common approach that speeds up the execution of all Six Sigma improvement projects.

This standardization creates a common language and a common cause among all employees. Many organizations implementing quality programs become mired in arguments and disagreements over methods and never move forward. A Six Sigma program, by contrast, focuses on reducing variability and reaching excellence

6.2.1 Program failure

Not surprisingly, the failures and limitations of Six Sigma are absent from all of the hype. The number one shortcoming is its failure to link Six Sigma projects to an overall guiding strategy. The most common points of failure are:

- Lack of executive management support

- Becomes only marketing tool

- Lack of well trained certified black belts and master black belts

- Lack of statistical knowledge

- Simplifying the statistical analysis and tools

- Fuzzy objectives.

- Too many objectives

- Organization politics

- Converting the existing well going projects to Six Sigma project to show the overall saving advantage

- Excessively broad scope.

- Defined or immeasurable metrics.

- No clear tie to financials.

- Disconnection from strategic or operating plan.

- Solution already identified

6.2.2 Cost of Poor Quality vs. Six Sigma:

Cost of poor quality has four elements:

- External cost: warranty claims, service cost

- Internal cost: the costs of labor, material associated with scrapped parts and rework

- Appraisal cost: these are materials for samples, test equipment, inspection labor cost, quality audits, inspection etc..

- Prevention cost: related to improving poor quality: quality planning, process planning, process control, and training.

Companies operating at three or four sigma usually spend between 25 percent of their revenues fixing problems. This is known as the cost of poor quality. Companies operating at Six Sigma typically spend less than 5 percent of their revenues fixing problems.

General Electric estimates that the gap between three or four sigma and Six Sigma was costing them between $8 billion and $12 billion per year.

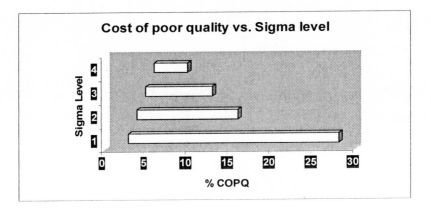

6.2.3 Structure the deployment of Six Sigma

All the executives from the CEO on down believed that Six Sigma needed to be an integral part of the company in order for it to succeed. That's why they had put the VPs in charge of coordinating the effort. But other elements of their deployment plans were inherently flawed from the perspective of integration. For example, the Black Belts, Master Black Belts, and official Six Sigma projects had no official relationship with any of the executives, division presidents, or others who were held accountable for financial performance. That automatically set up the Six Sigma effort in competition with the everyday work of the company.

6.2.4 Six Sigma leader selection

It is the fact that the knowledge of subject matter is the basic requirement for the success of any program. This fact applies to successful implementation of Six Sigma as well. Although the corporate Champion has spent most of her/his career in process improvement and problem solving, and was widely acknowledged as one of the best in those areas within the company, he/she lacked similar expertise in understanding statistics and statistical tools. He was therefore unable to influence effectively and create and maintain widespread commitment.

6.2.5 Organization readiness for Six Sigma Implementation

The implementation process of Six Sigma like any other new programs is different from organization to organization, but the preliminarily study must be conducted to evaluate the readiness of the organization for the program. The people directly involved with the Six Sigma implementation must receive extensive training and support. There are companies which have implemented the program to drive customer satisfaction, improve quality, and generate impressive financial payback. Yet for each of them there are hundreds of other companies who have achieved only a fraction of the full Six Sigma potential. Could it be that Six Sigma is in danger of ending up just like its predecessors? The question for every executive and Six Sigma support is "how good is *your* Six Sigma?" For many the answer will be "not as good as we hoped." Even those who are on the right track with Six Sigma can find ways to improve their deployment strategies and results. Organizations need to look at the warning signals of trouble, exposes common roadblocks that impede achievement of rapid results, and establishes a framework you can use to evaluate and improve your own Six Sigma initiative.

6.2.6 Result indicators of miss implementation

There are always some signs that indicate the program is not going well. Some of them such as: projects take longer and longer to complete• Financial returns drop steadily (Black Belts begin working on smaller projects that would be better suited to Green Belt. Project Champions do not fully support the projects, there is not a formal Black Belt certification program or down grading the certification program in order for employees to pass the tests. Black Belts and Champions have a difficult time recruiting people to staff projects Attendance at meetings drops off. More and more people question their involvement in Six Sigma ("it's taking time away from my regular job"; "this isn't helping me get more efficient, it's just adding to my workload"). Participation, commitment, and philosophy varies widely between locations and/or divisions • Instead of devoting high-caliber employees to Six Sigma deployment, "problem" individuals that others don't want are assigned to the Six Sigma team.

It is very important for the organization to integerad Six Sigma into their business strategies. The organization must evaluate the kind of results from the investment that they make on the program. Although at any Six Sigma presentation and introduction we can see the strategic integration of the program and operations, but in reality not too many of the executive apply the Six Sigma concept to their daily activities. One of the best ways to have successful implementation of the program is to build an "executive" version of the DMAIC (define - measure - analyze -improve - control) improvement strategy that lies at the heart of all Six Sigma processes. Perhaps the biggest advantage of DMAIC is the discipline it brings in helping us avoid the urge to jump into actions before we've defined, measured, and analyzed what is really going on. When applied to accelerating an existing Six Sigma initiative, the first step is an evaluation that helps you check on how well you have Defined the goals for Six Sigma, and Measure the impact of all your Six Sigma efforts.

6.2.7 Evaluation Strategy

Typically, an executive-level team oversees the evaluation of Six Sigma roadblocks. Depending on the scope of the effort, and their own comfort level with Six Sigma deployment, they may involve others in the organization or even outside experts in helping to conduct the evaluation and develop plans. The evaluation itself often has two main components: (1) in-depth interviews with a spectrum of people throughout the organization, and (2) a review of objectives and metrics that reflect Six Sigma performance. Having a mix of objective and subjective data is important. The objective data is needed so you can on firm whether the projects have had any measurable impact. In cases like these, however, subjective data gained from interviews is key to understanding the reasons for success or failure. (Survey data may also be informative, but not as helpful in probing for potential causes or problems).

6.3 The Economics of Six Sigma Quality

When companies embark on Six Sigma quality programs, what is their objective? Is it to reduce the process variance so that the half tolerance of the product characteristic is equal to six times the standard deviation? Or is it to have very few defects, say in the neighborhood of 50 to 100 per million? From the technical viewpoint. It might make sense to talk in terms of the process variance. From the managerial or customer viewpoint, the quality standards can be described in terms of defects per million. In addition, in many situations, adjusting the process to move the process average closer to the target value is relatively easier than improving the process to reduce the variance. Thus, if the goal is to reduce the number of defects, it does not make much sense to improve the process to Six Sigma levels and not center the process. Planning or allowing for the process average to draft 1.5 standard deviation from the target value just in case is similar to building up inventories when implementing just-in-time inventory management.

Of course, there might be processes in which the mean drifts away from the target value in an expected manner for example, due to tool wear. In such cases, the process average initially might be specifically set off-center to the target value to minimize the number of set-ups or tool changeovers. In these cases, the desired quality level (expressed by the number of defects per million) might be achieved through several combinations of off-centering and process standard deviations.

For example, a quality level of 3.4 defects per million parts can be achieved in the least three different ways:

- With 0.5 sigma off-centering with five sigma quality
- With 1 sigma off-centering with 5/5 sigma quality
- With 1.5 sigma off-centering with Six Sigma quality

How to achieve a specified quality level or a given number of defects per million depends on the costs associated with adjusting the process average vs. Reducing the process variance. If the process average can be centered and maintained at the target value, a four-sigma quality program results on only 63 defects per million, and even a 3.5-sigma program results in just 465 defects per million. If the process centering cannot be effectively controlled or monitored a little room could be allowed on each side of the specification to have some insurance against process shifts.

It is generally true that reducing the process variance involves such extensive efforts as continuous improvement programs, statistical design of experiments methodology, and capital investments is better technology. Alternatively adjusting the process to the target value might require less extensive efforts, such as employee training in statistical problem-solving methods and techniques. Certainly, companies want to reduce process variance, but being cost-effective is equally essential.

Quite a few company executives say they have embarked on Six Sigma quality programs. It is hoped that they know what they have embarked on and that they know how to process average significantly affects what Six Sigma quality really means.

It consists of the individual scorecards at the component, sub-assembly, and top-level assembly rolled up to the system level. As a first step, the number of opportunities at the parts, process, and performance levels is identified for each component being analyzed. An opportunity may be defined as a chance to do an operation successfully. The number of opportunities may also be considered an indicator of the complexity of the part, with a larger opportunity count indicative of a more complex part.

The occurrence of defects is characterized by the defects per unit (dpu). To avoid biasing the analysis by the complexity of the component, the dpu is normalized by the total number of opportunities to give the defects per million opportunities (dpmo). The dpmo is then related to the Sigma level through the normal distribution.

A typical reasonably well-controlled process is usually about Four Sigma (sigma being the standard deviation), or 6000 defects per million opportunities (dpmo). Thus, Six Sigma, which translates into a quality level of no more than 3.4 dpmo, represents a 2000 times improvement over "conventional wisdom."

Affordability is the primary motivation for the application of Six Sigma to the design, as it relates cost and value to each of the steps in the design scorecard. "Sometimes it can identify a change that can be made but it then has to answer the question 'why make it if it isn't going to impact cost?' We want to identify and attack the defects that have the largest impact on cost. There are a variety of ways we can do this—removing the part, adding automation, changing the process...."

To that end, Six Sigma is being linked to cost modeling work being done. Just as one-for-one part replacement of composites for traditional materials like metals is often not technologically sound, neither is it economically advantageous.

This turns out to be a fundamental problem. The major elements of cost which are: materials, labor, and overhead—can be estimated independently. But with composites, the processing options and manufacturing methods are so closely linked to design that these traditional costing methods are ineffective in discriminating among alternative designs. Allocated Based Accounting (ABC) shows promise as a tool for composites engineering design teams but only when it is employed during the early stages of design. Thus, it is extremely compatible with the application of Six Sigma to design. Future systems will require higher levels of product quality at lower cost and greater added value. As a result, the program has employed a number of tools to aid in creating the systems design, as well as to develop a database to satisfy future Army needs." The vision is that Six Sigma coupled to cost models will enable companies to priority rank the major sources of defect and cost. The information lets us focus on the most critical/high pay-off topics.

Another Key To Six Sigma Quality quoting begins, according to experts, with a better understanding of a customer's needs. For example "A quote just isn't a quote, it has to be a complete and thorough understanding of everything necessary to complete the parts perfectly, to spec and just in time, the organizations need to dig deeply to get all the facts possible on a book so we can evaluate alternative methods which can affect both price and quality. Organizations have to see beyond the numbers because prints can often show a part is greatly over-engineered.

Sales personnel, engineers and quoting department are trained to interpret prints and question what they see. They will make suggestions to improve a part, to make it less costly and less difficult to produce. Prints even come to us with errors. We have to be experienced enough to question what's given to us and get it corrected before we quote on it". Some organizations generated part drawings also help to improve quality in the quoting process. A CAD system drawing is easier to understand when quoting and far easier for shop personnel to follow when producing the part than hand-produced drawings. Even when sale is steadily increasing the number of books CAD produced and downloaded to the shop floor via computer network the human element is still as important as ever. It goes back to the original design and the original quote and quality of the input at that time. Customers, Suppliers Need Similar Inspection Systems.

With the demand for greater precision, and tools to achieve it such as statistical process control processes and equipment the human element for error still remains. Whoever knows for certain that the anvils of a micrometer are perfectly flat and parallel each time you use it?" This attitude underscores approach to quality-never take it for granted. Question everything.

Get the process right the first time prior to production and the need for inspection down. Companies need to make a strong case for vendor and customer agreeing to similar inspection methods and tools at the time a part book is quoted. An air gauge to check a hole diameter may give you a different reading than a dial bore gauge, plug gauge or Sunnen gauge. Measurement tools need to be the same and then we work with a customer to specify similar methods for determining quality at the time a book is quoted. It can be as simple as specifying gauging tools. Or it can be as complex as the computer and computer program that produces X Bar R charts for the entire statistical process control tracing of a part book"

Do Purchasing Departments Still Buy On Price? Or Quality? That's an unanswerable unending question, according to Jack Graeber, Vice President of Sales and Marketing at Northwest Swiss-Matic. "The point is, according to Graeber, our customer's purchasing people are measured by their ability to get good prices. So we have to find ways to achieve quality and meet the price requirements too. We quote twenty jobs to get one and that proves to us companies are shopping harder than ever for low prices. While we are constantly striving for perfect quality, we impose just as much discipline and pressure on ourselves to bring costs down. We prove every day with the quality we produce that better quality does not necessarily mean higher costs. In fact, it can actually lower production cost when flaws are eliminated and require no correction," Graeber added.

The Just-In-Time Bugaboo Northwest Swiss-Matic says that it also has to continue to find better ways to satisfy customers Just-In-Time delivery requirements. In the high volumes that it produces parts, Northwest Swiss-Matic points out that it may take 50,000 in process parts to produce and deliver 5,000 finished parts each day to a customer's dock for his JIT operations. "The questions of which pays for all of this become a major issue," reports Graeber, "as customers press us to hold more and more inventory while reducing theirs. The Golden Rule has to apply somewhere. Simpler, faster manufacturing processes are the never-ending goal in order to reduce in process parts and reduce lead times." Prompt Payment Equates With Prompt Deliveries "A lot is said about on-time deliveries but not too much about payment terms," reports Graeber. "Yet there are some OEM's who will beat on their vendors at both ends of the spectrum-for low cost, top quality parts, and delivered promptly. 'Men they won't pay for all of these services for at least 90 to 120 days. We think the Golden Rule has to apply more than ever-just-in-time part delivery and just-In-Time payment have to match up if we're to continue the process of constant improvement." "All business, good, or bad, starts with an attitude," states Martin. As complicated as it can get at times, good business still boils down to having the correct attitude and for sellers and buyers working together in the light spirit to solve a need. We feel we have that spirit and with our customers are on the way to world class status. It's an exciting time to be in this business. Many people have asked how we began the Six Sigma Quality program.

In 1985, a Motorola quality professional presented a paper describing the relationship of a product's early-life field reliability to the frequency of repair during the manufacturing process.

The conclusion was simply this - if during a product's manufacture you have to identify and fix defects incurred during its manufacture, you will miss defects that will affect the customer during the early life of the product. On the other hand, if your designs are robust and your manufacturing procedures are controlled so that virtually everything works right the first time, you are highly likely to ship products that will be free from failure in their early useful life.

Clearly, the objective is to eliminate the cause, not to identify and repair defects. We found later that this process also reduces defects in non-manufacturing operations such as order entry. In order to achieve the goal of "doing it right the first time," we need to established and communicate the process that we termed Six Sigma. Using the Six Sigma process has been critical to Motorola's success in its pursuit of continual quality improvement. Beyond this, because the pursuit of Six Sigma quality has required dramatic changes in processes and renewal of work flows, it has also driven new inventions.

Finally quality should not cost more. The whole process needs to be design to produce a high quality product with minimum cost on time. This will not happen unless the Six Sigma approach starts from the design of the product. Building the organization based on the Six Sigma standards will grantee an efficient process with high quality and to some extend low cost. But once again the cost if a factor that will be determine by complexity and importance factor.

6.4 Return on Six Sigma (ROSS) or Return on Investment (ROI)

Six Sigma is not only about quality in the traditional sense but also about helping the organization make more money and become profitable. The traditional quality, defined as conformance to internal requirements, has little to do with Six Sigma. Six Sigma is. To link this objective of Six Sigma with quality requires a new definition of quality.

Quality comes in two flavors:

Expected quality and *actual quality*. Expected quality is the known maximum possible value added per unit of input.

Actual quality is the current value added per unit of input. The difference between expected and actual quality is *waste*.

Six Sigma focuses on improving quality and reducing waste by helping organizations produce products and services better, faster and cheaper. In more traditional terms, Six Sigma focuses on defect prevention, cycle time reduction, and cost savings. Unlike mindless cost-cutting programs which reduce value and quality, Six Sigma identifies and eliminates costs which provide no value to customers, waste costs. For non-Six Sigma companies, these costs are often extremely high. Companies operating at three or four sigma typically spend between 25 and 40 percent of their revenues fixing problems. This is known as the cost of poor quality (COPQ).

Return on investment (ROI) or Return on Six Sigma (ROSS) helps companies turn over working capital faster, reduce capital spending, and make existing capacity available and new capacity unnecessary. It will add to the bottom line by reducing the unnecessary and non value adds activities.

"Six Sigma Advantage, Inc" has performed an analysis on the return on Six Sigma projects. They believe that "Even a small Six Sigma deployment can produce dramatic results. In our illustration below, deploying 15 Black Belts (to complete 3 projects each) and 15 Green Belts (to complete 2 projects each) will produce benefits of over $8,000,000 in the first year, typically at a cost of less the $2,000,000. Results like this are typical when Six Sigma is properly deployed".

Project Type	Average Project Savings	Number GB's / BB's Per Wave	Total Projects	Savings Year 1 (millions)
Black Belt	$150,000	15	45	$6.75
Green Belt	$ 50,000	15	30	$1.50
Totals	N/A	30	75	$8.25

Saving reports from the companies that have implemented Six Sigma

131

1. Motorola, has saved $15 billion during 11 years of Six Sigma discipline
2. AlliedSignal has had productivity gains of 6 percent in manufacturing in a two-year period
3. General Electric produced more than $2 billion in benefits in 1999, because of its Six Sigma efforts

There's mounting pressure on businesses to maintain a competitive edge in everything they do and to achieve complete customer and shareholder satisfaction. That means optimizing cycle time and equipment usage; having fewer rejects or errors; improving response time to customer inquiries; reducing inspection, maintenance, inventory, and other high costs; providing more employee development; and boosting financial results.

ROI is arguably the most popular metric when you need to compare the attractiveness of one business investment to another. The return on investment equals the present value of your accumulated net benefits (gross benefits less ongoing costs) over a certain time period divided by your initial costs. It is expressed as a percentage over a specific amount of time. If the a process or a technology obsoletes in three years, after three years. The equation for a 3-year ROI is:

(net benefit year 1 / (1+discount rate) + net benefit year 2 / (1+discount rate) + net benefit year 3 / (1+discount rate)) / initial cost.

So if the initial cost for your manufacturing company's small new software roll-out was $10,000, your annual benefits minus annual costs are constant at $5,000 for the next three years, and the discount rate is 10%, your 3-year ROI would be:

($5,000 / (1 + .1) + $5,000 / (1 + .1)^2 + $5,000 / (1 + .1)^3)/$10,000 = 124%

While ROI tells you what percentage return you will get over a specified period of time, it does not tell you anything about the magnitude of the project. So while a 124% return may seem initially attractive, would you rather have a 124% return on a $10,000 project or a 60% return on a $300,

6.4.1 Poor Project Estimates

The most important part of ROI calculations the accurate estimate of the nature of the project and duration time. Following are some of the key elements that need to be considered in order to create an accurate ROI:

- Unclear requirements
- Unclear parameters affecting estimates
- Little or no project databank to learn from
- Difficult to adjust the estimates once approved (we must plan to re-estimate)
- Estimates are rushed
- Inability to get acceptance of the estimates
- Imposed budgets, time and resources
- Lack of training/coaching on estimating techniques (and tools)

Finance and the project champion play a critical role in establishing the ROI factors. The financial numbers must be calculated according to the financial structure of the company.

The final numbers and also the calculated and expected ROI need to be approved by the fiancé group as well.

7   CHAPTER SEVEN

7.1 *Probability Distribution Functions*

Probability distributions functions are used to simulate and create a model for a behavior of data. The distribution function can also be used for prediction of the model and estimate the coefficient factors. Some practical uses of probability distributions are:

To calculate confidence intervals for parameters and to calculate critical regions for hypothesis tests. For univariate data, it is often useful to determine a reasonable distributional model for the data.

Probability distribution function $f(x)$, also describes the probability associated with events having (possibly multidimensional) coordinate x, must have these properties:

- It must be normalized so that the integral over the defined space gives unity:

$$\int_{-\infty}^{\infty} f(x)dx = 1 .$$

- It must be positive definite:

$$f(x) \geq 0 \quad \text{for all } x.$$

- The integral between specified limits must give the probability that events will occur in that interval:

$$\int_{a}^{b} f(x)dx = P(a < x < b) .$$

One often needs to transform such distribution functions to find probability distribution functions in terms of new variables. Consider the transformation from variable x to variable y where the relationship between y and x is known ($y(x)$). Then, if the probability distribution function $g(y)$ represents the same probability distribution function in terms of the variable y, to preserve the relationship of integrals to probability it must be true that

Statistical intervals and hypothesis tests are often based on specific distributional assumptions. Before computing an interval or test based on a distributional assumption, we need to verify that the assumption is justified for the given data set. In this case, the distribution does not need to be the best-fitting distribution for the data, but an adequate enough model so that the statistical technique yields valid conclusions.

Simulation studies with random numbers generated from using a specific probability.

There are two types of distributions:

7.2 *Continuous Distributions*

There are 27 continuous distributions. For the most part, they make use of three parameters: a location parameter, a; a scale parameter, b; and a shape parameter, c. there are a few exceptions to this notation. In the case of the normal distribution, for instance, it is customary to use $\propto$ for the location parameter and σ for the scale parameter. In the case of the beta distribution, there are two shape parameters and these are denoted by v and w.

Also, in some cases, it is more convenient for the user to select the interval via min and max than the location and scale. The location parameter merely shifts the position of the distribution on the x-axis without affecting the shape, and the scale parameter merely compresses or expands the distribution, also without affecting the shape. The shape parameter may have a small effect on the overall appearance, such as in the Weibull distribution, or it may have a profound effect, as in the beta distribution.

7.3 Discrete Distributions

There are nine discrete distributions. For the most part, they make use of the probability of an event, *p*, and the number of trials, *n*.

Probability distributions arise from experiments where the outcome is subject to chance. The nature of the experiment dictates which probability distributions may be appropriate for modeling the resulting random outcomes. There are two types of probability distributions - *continuous* and *discrete*.

Continuous (data)	**Continuous (statistics)**	**Discrete**
Beta	Chi-square	Binomia
Exponential	Noncentral Chi-square	Geometric
Uniformamma	F	NegativeBinomial
Lognormal	T	Hypergeometric
Weibull	Noncentral F	
Uniform	Noncentral t	
Normal		

"Table 4" distribution functions

7.4 Central limit theorem

From Wikipedia, the free encyclopedia:

Central limit theorems are a set of weak convergence results in probability theory. Intuitively, they all express the fact that any sum of many small independent random variables is approximately normally distributed. These results explain the ubiquity of the normal distribution.

The most important and famous result is simply called *The Central Limit Theorem*; it is concerned with independent variables with identical distribution whose expected value and variance are finite. Several generalizations exist which do not require identical distribution but incorporate some condition, which guarantees that none of the variables exert a much larger influence than the others. Two such conditions are the *Lindeberg condition* and the *Lyapunov condition*. Other generalizations even allow some "weak" dependence of the random variables.

Let $X_1, X_2, X_3, \ldots$ be a sequence of random variables which are defined on the same probability space, share the same probability distribution D and are independent. Assume that both the expected value μ and the standard deviation σ of D exist and are finite.

Consider the sum $: S_n = X_1 + \ldots + X_n$. Then the expected value of S_n is $n\mu$ and its standard deviation is $\sigma\, n^{1/2}$. Furthermore, the distribution of S_n approaches the normal distribution $N(n\mu, \sigma^2 n)$ as n approaches ∞.

In order to clarify the word "approaches" in the last sentence, we normalize S_n by setting

$Z_n = (S_n - n\mu) / (\sigma\, n^{1/2})$.

Then the distribution of Z_n converges towards the standard normal distribution $N(0,1)$ as n approaches ∞. This means: if $F(z)$ is the cumulative distribution function of $N(0,1)$, then for every real number z, we have:

$\lim_{n \to \infty} \Pr(Z_n \leq z) = F(z)$.

If the third central moment $E((X_1-\mu)^3)$ exists and is finite, then the above convergence is uniform and the speed of convergence is at least on the order of $1/n^{1/2}$).

137

An equivalent formulation of this limit theorem starts with $A_n = (X_1 + \ldots + X_n) / n$ which can be interpreted as the mean of a random sample of size n. The expected value of A_n is μ and the standard deviation is $\sigma / n^{\frac{1}{2}}$. If we normalize A_n by setting $Z_n = (A_n - \mu) / (\sigma / n^{\frac{1}{2}})$, we obtain the same variable Z_n as above, and it approaches a standard normal distribution.

Note the following "paradox": by adding many independent identically distributed *positive* variables, one gets approximately a normal distribution. But for every normally distributed variable, the probability that it is negative is non-zero! How is it possible to get negative numbers from adding only positives? The key lies in the word "approximately". The sum of positive variables is of course always positive, but it is very well approximated by a normal variable (which indeed has a very tiny probability of being negative).

More precisely: the fact that, for every n there is a z such that $\Pr(Z_n \leq z) = 0$ does not contradict that for every z we have $\lim_{n \to \infty} \Pr(Z_n \leq z) > 0$, because in the first case z may depend on n and in the second case n is increased for a fixed z.

Following are some of the probability distribution functions that are used widely:

7.5 *Normal distribution*

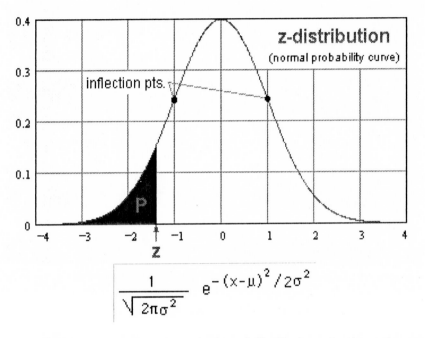

$$\frac{1}{\sqrt{2\pi\sigma^2}}\ e^{-(x-\mu)^2/2\sigma^2}$$

Normal distributions are a family of distributions that have the same general shape. They are symmetric with scores more concentrated in the middle than in the tails. Normal distributions are sometimes described as bell shaped. Examples of normal distributions are shown below. Notice that they differ in how spread out they are. The area under each curve is the same. The height of a normal distribution can be specified mathematically in terms of two parameters the mean (μ) and the standard divination (σ).

"Figure 11, Normal distribution Function"

When a coin is flipped, the outcome is either a head or a tail; when a magician guesses the card selected from a deck, the magician can either be correct or incorrect; when a baby is born, the baby is either born in the month of March or is not. In each of these examples, an event has two mutually exclusive possible outcomes. For convenience, one of the outcomes can be labeled "success" and the other outcome "failure." If an event occurs N times (for example, a coin is flipped N times), then the binomial distribution can be used to determine the probability of obtaining exactly r successes in the N outcomes. The binomial probability for obtaining r successes in N trials is: where P(r) is the probability of exactly r successes, N is the number of events, and π is the probability of success on any one trial. This formula assumes that the events:

(a) are dichotomous (fall into only two categories)

(b) are mutually exclusive

(c) are independent and

(d) are randomly selected

7.6 T Student Distribution

The Student t distribution Student T Distribution is the distribution followed by the ratio of a variable that follows the normal distribution to the square root of one that follows the chi-square distribution with degrees of freedom. The distribution characterizes the uncertainty in a mean when both the mean and variance are obtained from data.

$$\Phi_t(t,\nu) = \frac{\Gamma\left(\frac{\nu+1}{2}\right)}{\sqrt{\pi\nu}\Gamma\left(\frac{\nu}{2}\right)}\left(1+\frac{t^2}{\nu}\right)^{-\frac{\nu+1}{2}}$$

"Figure 12, T student distribution"

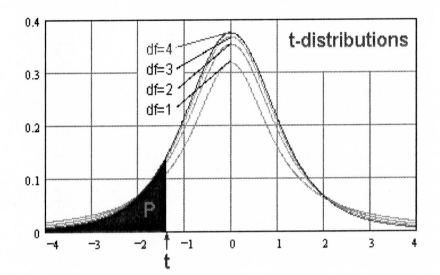

7.7 *Chi-Square distribution*

The chi-square distribution Chi Square Distribution is the distribution of the quantity, where the are random variables which follow a normal distribution with mean zero and unit variance. The chi-square distribution gives the distribution of variances of samples from a normal distribution.

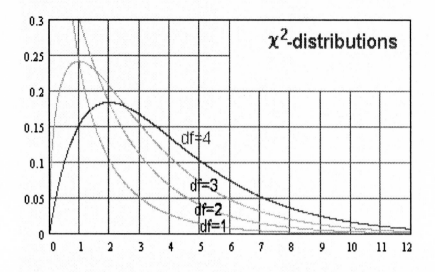

Figure 13, Chi square distribution

7.7.1.1 *The Gamma Distributions*

Functions can be classified as algebraic and transcendental. An algebraic function is a function that is a root of a polynomial equation. A function that is not a root of a polynomial equation is called transcendental. Most of the functions that describe natural phenomena turn out to be transcendental functions as are the trigonometric, logarithmic, exponential, and hyperbolic functions.

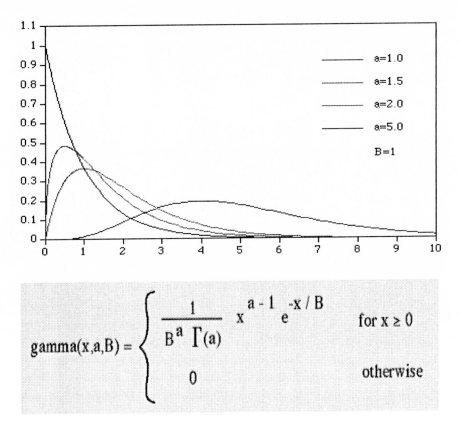

$$gamma(x,a,B) = \begin{cases} \dfrac{1}{B^a \; \Gamma(a)} \; x^{a-1} \; e^{-x/B} & \text{for } x \geq 0 \\ \\ 0 & \text{otherwise} \end{cases}$$

Figure 14, Gamma distribution

The theory of higher transcendental functions was elaborated by Euler, (1707-1783) who also introduced the beta and gamma transcendental functions. Most sampling distributions of inferential statistics belong to the family of the gamma density functions. Some textbooks on statistics ascribe the t-distribution to Student and the F distribution to Snedecor. These statisticians only called the attention to the applicability of some of the higher transcendental function to the theory of statistical inference. However, the gamma density functions are due to Euler. These functions have a general form.

7.8 The F Distribution

Among the higher transcendental functions, a frequently used function within the area of statistical inference is the inverted beta distribution, also called, as coined by Snedecor, the F distribution. As other probability distributions, the F distribution belongs to the family of gamma functions. The density function for the F-distribution, associated with certain number of degrees of freedoms signified by the Greek letter is:

$$y = \left[\frac{\Gamma\left(\frac{v_1 + v_2}{2}\right)}{\Gamma\left(\frac{v_1}{2}\right) + \Gamma\left(\frac{v_2}{2}\right)} \right] \left(\frac{v_1}{v_2}\right)^{v_1/2} \left(x^{(v_1/2)-1}\right) \left(1 + \frac{v_1 x}{v_2}\right)^{-(v_1+v_2)/2}$$

Figure 15, F distribution

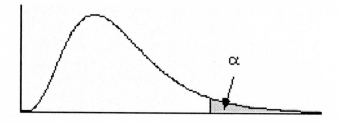

7.9 Binomial Equation

The binomial distribution describes the possible number of times that a particular event will occur in a sequence of observations. The binomial distribution is used when a researcher is interested in the occurrence of an event, not in its magnitude. For instance, in a clinical trial, a patient may survive or die. The researcher studies the number of survivors, and not how long the patient survives after treatment.

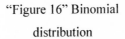

"Figure 16" Binomial

distribution

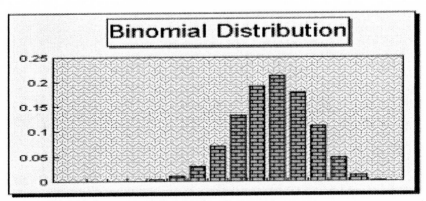

Another example is whether a person is ambitious or not. Here, the binomial distribution describes the number of ambitious persons, and not how ambitious they are. The fundamental distribution of the general linear model of data analysis is the binomial distribution. The equation for the binomial distribution is:

$$P(y) = \frac{n!}{y!(n-y)!} \pi^{y}(1-\pi)^{n-y}$$

7.10 The Poisson distribution

The Poisson distribution may be derived directly as the probability of a rare event in a large number of trials, or else it may be derived as a limiting case of the binomial distribution. We shall begin by taking the latter approach.

$$\Pr(X = x) = \frac{\lambda^{x} e^{-\lambda}}{x!}, \quad x = 0, 1, \ldots, \infty$$

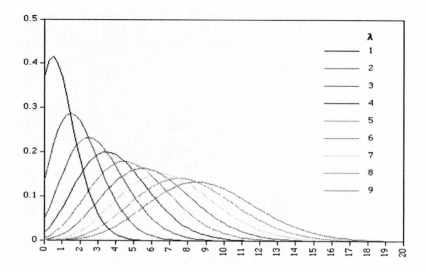

"Figure 17" Poisson distribution

7.11 Exponential distribution

The **exponential distribution** is a continuous probability distribution with probability density function:

p(t) = 0 for t<0

p(t) = exp(-t/λ)/λ for t ≥ 0

where λ > 0 is a parameter of the distribution. The distribution is useful in a situation where an object is initially in state A and can change to state B with constant probability per unit time, equal to 1/λ. Such a process is called a Poisson process. A random variable following the exponential distribution describes the time at which the state switches. Therefore, the integral from 0 to T over p is the probability that at time T the object is in state B.

$$\text{exponential}(x,\lambda) = \begin{cases} \lambda\, e^{-\lambda x} & x \geq 0, \lambda > 0 \\ 0 & \text{otherwise} \end{cases}$$

Figure "18" Exponential distribution

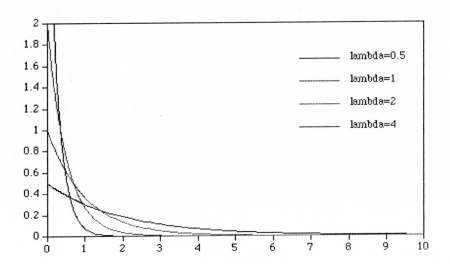

8   CHAPTER EIGHT

8.1   *Implementation of Six Sigma*

Although most of the executive managers might believe that implementing Six Sigma will resolve their problems and makes the company super profitable, it is so important to implement it correctly and not using the name as a sales tool. A good Six Sigma business strategy involves the measurement of how well business processes meet their objectives and offers strategies to make needed improvements. The application of the techniques to all function results in a very high level of quality at reduced costs with a reduction in cycle time, resulting in improved profitability and a competitive advantage. It needs to be emphasized that organizations do not need to use all the measurement units that might be associated with Six Sigma. It is most important to choose the best set of measurements for their situation and focus their emphasis on the *wise* integration of statistical and other improvement tools.

A good Six Sigma implementation plan defines Six Sigma projects in critical areas of the business. A road map for selected projects involving the phases of measure, analyze, improve, and control is described in Breyfogle (1999) as a Smarter Six Sigma Solutions (S^4) approach.

Once an implementation plan is in place, the issue of deployment comes to the fore.

After over two decades of experience with quality improvement, there is now a solid body of scientific research regarding the experience of thousands of companies implementing major programs such as Six Sigma. Researchers have found that successful deployment of Six Sigma involves focusing on a small number of high-leverage items. The steps required to successfully implement Six Sigma are well-documented.

8.2 Leadership training

Start by providing senior leadership with training in the principles and tools they need to prepare their organization for success. Using their newly acquired knowledge, senior leaders direct the development of a management infrastructure to support Six Sigma. Simultaneously. This involves reducing levels of organizational hierarchy, removing procedural barriers to experimentation and change, and a variety of other changes designed to make it easier to try new things without fear of reprisal.

8.3 Close communication system

Any quality system is developed to establish a close loop communication with customer, employees, and suppliers. This includes developing *rigorous* methods of obtaining and evaluating customer, employee and supplier input. Base line studies are conducted to determine the starting point and to identify cultural, policy, and procedural obstacles to success.

8.4 Successful Training program

Remedial skills education is provided to assure that adequate levels of literacy and numeracy are possessed by all employees. Top-to-bottom training is conducted in systems improvement tools, techniques, and philosophies.

8.5 Framework for continuous improvement

A continues process improvement is developed, along with a system of indicators for monitoring progress and success. Six Sigma metrics focus on the organization's strategic goals, drivers, and key business processes.

8.6 Project Selection

One of the key success of a Six Sigma program in any organization is the correctly selection of projects. Business processes to be improved are chosen by management, and by people with intimate process knowledge at all levels of the organization. Six Sigma projects are conducted to improve business performance linked to measurable financial results. This requires knowledge of the organization's constraints. Organization must develop a criteria for how to select a project. Projects with Fuzzy objectives or Too many objectives will not be a good candidate for a Six Sigma program.

8.7 Project Management and Team

Six Sigma projects are conducted by individual employees and teams led by Green Belts and assisted by Black Belts.

Although the approach is simple, it is by no means easy. But the results justify the effort expended. Research has shown that firms that successfully implement Six Sigma perform better in virtually every business category, including return on sales, return on investment, employment growth, and share price increase.

8.8 Champion Training

As mentioned in the past the executive commitment to program implementation is the key to the success of Six Sigma. That's the reason why implementing Six Sigma begins with the training of executive leaders. It is not enough for executives to support Six Sigma, they must lead the strategy. Senior managers who write memos on the importance of quality but still drive through volume-based metrics will not have success with projects to achieve bottom-line benefits and improve quality. A project to increase quality in this organization will not be accomplished if volume is the only measure and rewarded accordingly. What is measured and rewarded drives employee behavior.

The senior management leadership was often a typical missing element for success with past Total Quality Management (TQM) initiatives. Projects were not typically selected from a strategic, executive perspective. Effective usage of statistical tools often did not get recognized and the overall company culture was not impacted. For true success with Six Sigma, executive level leadership is needed that asks the right questions leading to the wise application of statistical tools and other Six Sigma methodologies across organizational boundaries.

A question we frequently hear from executives is "How does Six Sigma fit with other corporate initiatives?" The response is that Six Sigma should not be considered just another initiative but should integrate other programs as an overall business strategy at the executive level.

High-level control charts can be used to identify common cause issues and help re-focus firefighting activity to process improvement efforts. These charts generate "30,000 foot level" metrics that can be used to track progress in key areas or Key Process Output Variables (KPOV's) of the business at the executive level, giving insight into what areas of the business should be given focus relative to project selection. Once projects are defined, the tool/strategy that is most appropriate for the given situation (e.g., lean manufacturing, Kaizen, Six Sigma Design of Experiments, or Six Sigma Gage R&R) can then be determined.

8.9 Importance of Customer Quality Index

Establishing a Customer Focus mindset within an organization goes hand in hand with Senior Management Leadership when creating a successful Six Sigma business strategy. The factors that are critical to your customers' success are necessary to a process improvement team's true success. Therefore, evaluating customers' perception of quality should be at the forefront of the implementation process.

Every complaint from a customer should be viewed as an opportunity for growth and increased market share, a spotlight on areas needing process improvement focus. The key to success in this initial step is to make it easy for your customers' comments to be heard. Accurately capturing the "voice of the customer" is a laborious process that is frequently skipped over. Organizations believe they understand what is important to their customers but are frequently surprised when they actually spend the time to quantify their actual needs.

Depending on the size of an organization and its core values, the word customer can take on many different definitions. When collecting feedback, care should be taken to include a comprehensive view of customers. By combining external feedback with such things as internal business strategies, employee needs and government regulations, an organization will obtain a balanced list of customer needs.

Learning through customer feedback of what works and what does not, will help to establish a mindset of continual process improvement within your organization.

Jack Welch, CEO of GE and the most visible advocate of Six Sigma, himself has been quoted to say that a business strategy alone will not generate higher quality throughout an organization.

8.10   Organizations' Strategic Goals

The organization needs to set smart goals, which are simple, measurable, reasonable and time based. Six Sigma must be viewed as a method to meet strategic goals, these goals need to be measurable and have the focus of executive management. Asking the right question means defining the strategic goals of your organization. It requires communicating to your employees what is strategic and why and following up those statements with executive focus and metrics.

Again, Six Sigma should not replace existing organizational initiatives, but instead create an infrastructure that offers a tactical approach to determine the best solution for a given process/situation. There has to be accountability. There must be enthusiasm.

It is similar to a self-help program. What you put into it is what you will get out of it. If you pay it "lip service," you will get mediocre results. If it is utilized as a business strategy, it becomes a focused approach to meeting the strategic goals defined by executive management, allowing the application of resources in critical areas to the bottom-line.

8.11 Cultural Challenges

Many companies attempt to improve products with numerous small changes or "tweaks" to their current processes; however, changes are frequently not documented and the associated results not reported. Substantial results are rarely obtained with this halfhearted method of change. When employees in this type of corporate culture hear of a new initiative such as Six Sigma, they wonder what will be different.

As a program or initiative, Six Sigma risks becoming the "flavor of the month" and will not capture the buy in necessary to reap a large return on the investment in training. With this approach, employees may end up viewing Six Sigma as a program similar to Total Quality Management (TQM) and other quality "programs", which may have experienced limited success within their organization.

In today's constantly changing market place companies that are able to embrace change in a focused and proactive manner are leaders in their field. Companies who not only master the technical side of Six Sigma but also overcome the cultural challenges associated with change can realize significant bottom-line benefits.

Companies are embracing Six Sigma not only to reduce defects, but also as a catalyst to change the culture of their company, impacting how employees engage in their every day work. Every company that takes on Six Sigma, undergoes a unique journey of integrating the methodology into their current culture. Infrastructures vary significantly between organizations, depending upon their distinct culture and strategic business goals. Launching a Six Sigma business strategy is an excellent opportunity to assess current culture in an organization. Consider the following questions:

- How has your company historically dealt with change initiatives?

- Does your company make consistent changes that don't last?

- How effective are your project teams?

- Are you frequently focusing on the same problem?

- How do our employees attack problems and conduct their daily work?

- What is required within your company culture to make continual process improvement a lasting change?

- What will prevent your company from achieving success with Six Sigma?

A Force Field Diagram can be created to facilitate the understanding of how well Six Sigma integrates with the current culture in an organization. By evaluating and weighting the key cultural drivers and restraints to embracing Six Sigma, organizations can develop action plans that enhance the key drivers and mitigate the critical restraints.

Organizations need to have a direction that ingrains a process-focused, proactive mindset into the way all your employees approach their every day work. When successful, Six Sigma becomes part of your culture.

8.12 Define Customer /Internal Metrics

Lack of statistical techniques is often causing the organization to misinterpret the result and make a wrong decision. It has been observed that the management over simplifies the statistical tools in order to sale the program to their staff. Unfortunately that has created a false and negative impact on the company's bottom line results and often has been blamed on the Six Sigma being expensive to implement. There is no "one size fits all" metric applicable to every Six Sigma project. Effective metrics are cross-functional, providing a holistic view of the process and contributing insight to the project team. Many resources can be wasted if Six Sigma metrics are not applied wisely and subsequently used to orchestrate improvement activities, "fire prevention" as opposed to "fire fighting", as mentioned previously. Unfortunately, much confusion exists relative to the metrics of Six Sigma.

8.13 Sigma quality level

A basic goal of a Six Sigma program might be to produce at least 99.99966% "quality" at the "process step" or part level within an assembly (i.e., no more than 3.4 defects per million parts or process steps if the process mean were to shift by as much as 1.5). If, for example, there was on the average 1 defect for an assembly that contained 40 parts and 4 process steps, practitioners might consider that the assembly would be at a four sigma quality level, since the number of defects in parts per million is: $(1/160)(10^6)=62500$.

A sigma quality level metric can be deceiving. Determining the number of opportunities for any given process can be dramatically different between individuals. For example, one process might have a 50 percent defective unit rate and a sigma quality level much greater than six, while another process might have a .01 percent defective unit rate and have a sigma quality level much worse than six. To illustrate this, first consider the counting of opportunities for failure within a computer chip as junctions and "components." The sigma quality level metric for this situation typically leads to a very large number of opportunities for failure for a given computer chip; hence, a very high sigma quality level is possible even when the defective rate per unit is high. Compare this situation to another situation where there were only a very few number of components or steps required for a process.

The sigma quality level metric for this situation typically leads to a very low number of opportunities for failure; hence, a very low sigma quality level metric is possible even when the defective rate per unit is low.

The sigma quality level includes a ± 1.5 value to account for "typical" shifts and drifts of the mean, where is the standard deviation of the process. This sigma quality level relationship is not linear. In other words a percentage unit improvement in parts per million (PPM) defect rate (or defect per million opportunity (DPMO)) rate does not equate to the same percentage improvement in the sigma quality level; the improvement from 4.1 to 4.2 sigma quality level is not the same as improvement from 5.1 to 5.2 sigma quality level.

The sigma quality level metric can only be determined when there are specifications. Service/transactional applications do not typical have specifications like manufacturing does. When a sigma quality level is forced within a service/transactional situation this can lead to the fabrication of specifications and alterations of these "specifications" to make the "numbers look good."

Process capability index is another Six Sigma metric that is used to describe. It means how well a process meets requirements. A Six Sigma quality level process is said to translate to process capability index values for Cp and Cpk requirement of 2.0 and 1.5 respectively. Unfortunately there is also much misunderstanding with this metric, even though the following basic equations for these metrics are simple.

Computer programs often will not even give the same process capability answer for a given set of data. Some programs consider the standard deviation to be short-term, while others consider standard deviation to be long-term. Implement Six Sigma describes eight different approaches that could be used to determine standard deviation.

A wise approach to implementing Six Sigma is not to force a sigma quality metric within the various groups and/or projects within an organization. It is most important to use the right metric for any given situation. It is essential that the sigma quality level metric be included, along with the other Six Sigma metrics, within all Six Sigma training. The positive, negative, and controversial aspects of each Six Sigma metric should be covered within the training so that organizations can more effectively communicate with their customers and suppliers.

Often customers and suppliers ask the "wrong question" relative to Six Sigma and other metrics. When an organization understands the plusses and minuses of each metric they can work with their customers and/or suppliers to direct their efforts toward the best metric for a given situation rather than reacting to issues that result from "mandated" metrics that makes no sense.

Care must be taken that the training an organization receives in Six Sigma metrics is not avoided. In addition to the wise selection of metrics, Six Sigma training should also address the wise use of statistical methodologies, providing insight to how one can best determine what is truly causing a problem.

8.14 Project Execution

It is obvious that there is much upfront work required to establish a support infrastructure and strategic goals before project work begins. It is the job of executives and the steering committee to integrate the voice of the customer into the strategic goals of the organization. Much work is done before projects are even started to transform comprehensive customer feedback and internal business goals into strategic Six Sigma goals. Goals without a roadmap can be detrimental.

8.15 Conclusion

Six Sigma is a great foundation for the quality system in an organization. It is a long-term commitment. Treating deployment as a process allows objective analysis of all aspects of the process, including project selection and scooping. Projects should be selected that meet the goals of an organization's business strategy. Six Sigma can then be utilized as a roadmap to effectively meet those goals.

Utilizing lessons learned and incorporating them into subsequent waves of an implementation plan creates a closed feedback loop and real opportunities for improvement. Deploying Six Sigma through projects can lead to dramatic bottom line benefits if the organization invests the time and executive energy necessary to implement a process to create a successful Six Sigma infrastructure.

Creating and implementing Six Sigma does not guarantee tangible benefits within an organization. However, when Six Sigma is implemented wisely as a business strategy accompanied with effective metrics, as summarized mentioned, organizations can achieve significant bottom-line benefits. Through the wise implementation of Six Sigma, the success of individual projects can build upon each other gaining the sustained attention of executive management and resulting in a corporate culture change from a reactive or fire-fighting environment to a learning organization, understanding threats and recognizing new opportunities for growth, not only to survive but to actually thrive within competitive environments.

9 CHAPTER NINE

9.1 *Continuous improvement methods*

There are lots of methods that the organization can implement and achieve a good result in terms of quality, productivity, customer satisfaction and profitability. Some of these methods have been around for years and organizations have had success and failure by implementing or miss implementation of them. It is obvious the success and failure of any such programs highly depends on the vision, commitment, leadership and philosophy of the organization. Organizations who try to implement these programs without the executive management full support and commitment always face chaise and drop the program at the beginning stages. Among these method lean manufacturing, 5S and Kaizen are very popular and have been used by organizations around the world.

9.2 *Lean Manufacturing*

It can be defined as a systematic approach to identifying and eliminating waste (non-value-added activities) through continuous improvement by flowing the product at the pull of the customer in pursuit of perfection. Companies today are in a constant battle for the bottom line. Simply stated you either make money or you're out of business! Most companies starting the Lean journey don't finish. Tom Peters in his book "The Circle of Innovation" stated it correctly when he said "it's easier to kill a company than to change it." This is exactly the problem! Lean manufacturing takes disciple and team work. Without these in place it is impossible to make the system stick.

9.2.1 Value

In lean production, the value of a product is defined solely by the customer. The product must meet the customer's needs at both a specific time and price. The thousands of mundane and sophisticated things that manufacturers do to deliver a product are generally of little interest to customers. To view value from the eyes of the customer requires most companies to undergo comprehensive analysis of all their business processes. Identifying the value in lean production means to understand all the activities required to produce a specific product, and then to optimize the whole process from the view of the customer. This viewpoint is critically important because it helps identify activities that clearly add value, activities that add no value but cannot be avoided, and activities that add no value and can be avoided.

9.2.2 Continuous Improvement

The transition to a lean environment does not occur overnight. A continuous improvement mentality is necessary to reach your company's goals. The term "continuous improvement" means incremental improvement of products, processes, or services over time, with the goal of reducing waste to improve workplace functionality, customer service, or product performance (Suzaki, 1987). Continuous improvement principles, as practiced by the most devoted manufacturers, result in astonishing improvements in performance that competitors find nearly impossible to achieve.

Lean production, applied correctly, results in the ability of an organization to learn. As in any organization, mistakes will always be made. However, mistakes are not usually repeated because this is a form of waste that the lean production philosophy and its methods seek to eliminate.

9.2.3 Customer Focus

A lean manufacturing enterprise thinks more about its customers than it does about running machines fast to absorb labor and overhead. Ensuring customer input and feedback assures quality and customer satisfaction, all of which support sales.

9.2.4 Perfection

The concept of perfection in lean production means that there are endless opportunities for improving the utilization of all types of assets. The systematic elimination of waste will reduce the costs of operating the extended enterprise and fulfills customer's desire for maximum value at the lowest price. While perfection may never be achieved, its pursuit is a goal worth striving for because it helps maintain constant vigilance against wasteful practices.

9.2.5 Focus on Waste

The aim of Lean Manufacturing is the elimination of waste in every area of production including customer relations, product design, supplier networks, and factory management. Its goal is to incorporate less human effort, less inventory, less time to develop products, and less space to become highly responsive to customer demand while producing top quality products in the most efficient and economical manner possible.

Essentially, a "waste" is anything that the customer is not willing to pay for. Typically the types of waste considered in a lean manufacturing system include:

9.3 Overproduction

To produce more than demanded or produce it before it is needed. It is visible as storage of material. It is the result of producing to speculative demand.

Overproduction means making more than is required by the next process, making earlier than is required by the next process, or making faster than is required by the next process. Causes for overproduction waste include:

- Just-in-case logic

- Misuse of automation

- Long process setup

- Unleveled scheduling

- Unbalanced work load

- Over engineered

- Redundant inspections

9.3.1 Waiting

For a machine to process should be eliminated. The principle is to maximize the utilization/efficiency of the worker instead of maximizing the utilization of the machines. Causes of waiting waste include:

- Unbalanced work load

- Unplanned maintenance

- Long process set-up times

- Misuses of automation

- Upstream quality problems

- Unleveled scheduling

9.3.2 Inventory or Work in Process (WIP)

Is material between operations due to large lot production or processes with long cycle times. Causes of excess inventory include:

Protecting the company from inefficiencies and unexpected problems

- Product complexity

- Unleveled scheduling

- Poor market forecast

- Unbalanced workload

- Unreliable shipments by suppliers

- Misunderstood communications

- Reward systems

- Less inventory required throughout the production process, raw material, WIP, and finished goods.

9.3.3 Processing waste

Should be minimized by asking why a specific processing step is needed and why a specific product is produced. All unnecessary processing steps should be eliminated. Causes for processing waste include:

- Product changes without process changes

- Just-in-case logic

- True customer requirements undefined

- Over processing to accommodate downtime

- Lack of communications

- Redundant approvals

- Extra copies/excessive information

9.3.4 Transportation

Transportation does not add any value to the product. Instead of improving the transportation, it should be minimized or eliminated (e.g. forming cells). Causes of transportation waste include:

- Poor plant layout

- Poor understanding of the process flow for production

- Large batch sizes, long lead times, and large storage areas

9.3.5 Motion

Motion of the workers, machines, and transport (e.g. due to the inappropriate location of tools and parts) is waste. Instead of automating wasted motion, the operation itself should be improved. Causes of motion waste include:

- Poor people/machine effectiveness

- Inconsistent work methods

- Unfavorable facility or cell layout

- Poor workplace organization and housekeeping

- Extra "busy" movements while waiting

9.3.6 Making defective products

Poor quality is a pure waste. Prevent the occurrence of defects instead of finding and repairing defects. Causes of processing waste include:

- Weak process control

- Poor quality

- Unbalanced inventory level

- Deficient planned maintenance

- Inadequate education/training/work instructions

- Product design

- Customer needs not understood

9.3.7 Underutilizing people

Underutilizing people or not taking advantage of people's abilities. Causes of people waste include:

- Old guard thinking, politics, the business culture

- Poor hiring practices

- Low or no investment in training

- Low pay, high turnover strategy

Nearly every waste in the production process can fit into at least one of these categories. Those that understand the concept deeply view waste as the singular enemy that greatly limits business performance and threatens prosperity unless it is relentlessly eliminated over time. Lean manufacturing is an approach that eliminates waste by reducing costs in the overall production process, in operations within that process, and in the utilization of production labor. The focus is on making the entire process flow, not the improvement of one or more individual operations.

Prepare and motivate people and widespread orientation to Continuous Improvement, quality, training and recruiting workers with appropriate skills and create common understanding of need to change to lean.

9.3.8 Employee involvement

Push decision making and system development down to the "lowest levels"

Trained and truly empowered people and share information and manage expectations Identify and empower champions, particularly operations managers. Remove roadblocks (i.e. people, layout, systems). Make it both directive yet empowering

9.3.9 Atmosphere of experimentation

Tolerating mistakes, patience, etc. Willingness to take risks

9.4 KAIZEN

KAIZEN is a Japanese word meaning gradual and orderly, continuous improvement. The KAIZEN business strategy involves everyone in an organization working together to make improvements 'without large capital investments'. Kaizen is a culture of sustained continuous improvement focusing on eliminating waste in all systems and processes of an organization. The kaizen strategy begins and ends with people. With kaizen, an involved leadership guides people to continuously improve their ability to meet expectations of high quality, low cost, and on-time delivery. Kaizen transforms companies into 'Superior Global Competitors'.

9.4.1 Two Elements of KAIZEN

There are two elements that construct KAIZEN, improvement/change for the better and ongoing/continuity. Lacking one of those elements would not be considered kaizen. For instance, the expression of "business as usual" contains the element of continuity without improvement. On the other hand, the expression of "breakthrough" contains the element of change or improvement without continuity. Kaizen contains both.

9.5 5-S

The 5-S technique is another continues improvement tool that has been widely practiced by quality organizations world-wide. Most 5-S practitioners consider 5-S useful not just for improving their physical environment, but also for improving their thinking processes too. Many of the everyday problems could be solved through adoption of this practice.

The 5-S consists of:

1. Seiri (Organization)

2. Seiton (Neatness)

3. Seiso (Cleaning)

4. Seiketsu (Standardizing)

5. Shtshke (Discipline)

9.5.1 Seiri (Organization)

Commonly used tools should be readily available and everything in a work area for the most efficient and effective retrieval and return to its proper place. Storage areas, cabinets and shelves should be properly labeled. Paint floors to make it easier to spot dirt, waste materials and dropped parts and tools. Outline areas on the floor to identify work areas, storage areas, finished product areas, etc. Put shadows on tool boards, making it easy to quickly see where each tool belongs. In an office, provide bookshelves for frequently used manuals, books and catalogs. Labels the shelves and books so that they are easy to identify and return to their proper place.

9.5.2 Seiton (Neatness)

Keep only what is necessary. Materials, tools, equipment and supplies that are not frequently used should be moved to a separate, common storage area. Items that are not used should be discarded. Don't keep things around just because they might be used, someday. Sorting is the first step in making a work area tidy. It makes it easier to find the things you need and frees up additional space. As a result of the sorting process you will eliminate (or repair) broken equipment and tools. Obsolete fixtures, molds, jigs, scrap material, waste and other unused items and materials are disposed of.

9.5.3 Seiso (Cleaning)

Once you have everything, from each individual work area up to your entire facility, sorted (cleaned up) and organized, you need to keep it that way. This requires regular cleaning, or to go along with our third S, "shining" things up regular, usually daily, cleaning is needed or everything will return to the way it was. This could also be thought of as inspecting. While cleaning it's easy to also inspect the machines, tools, equipment and supplies you work with. Problems can be identified and fixed when they are small. If these minor problems are not addressed while small, they could lead to equipment failure, unplanned outages or long - unproductive - waits while new supplies are delivered. When done on a regular, frequent basis, cleaning and inspecting generally will not take a lot of time, and in the long run will most likely save time.

9.5.4 Seiketsu (Standarding)

the fourth step is to simplify and standardize. The good practices developed in steps 1 through 3 should be standardized and made easy to accomplish. Develop a work structure that will support the new practices and make them into habits. As you learn more, update and modify the standards to make the process simpler and easier. **One of the hardest steps** is avoiding old work habits.

It's easy to slip back into what you've been doing for years. That's what everyone is familiar with. It feels comfortable. Use standards to help people work into new habits that are a part of your Five S program. Any easy way to make people aware of, and remind them about the standards is to use labels, signs, posters and banners. For example, use a Poster Printer to create large format signs, posters and banners.

9.5.5 Shitshke (Decipline)

Have a formal system for monitoring the results of your Five S program. Don't expect that you can clean up, get things organized and labeled, and ask people to clean and inspect their areas every day — and then have everything continue to happen without any follow-up. Continue to educate people about maintaining standards. What there are changes - such as new equipment, new products, new work rules - that will effect your Five S program and adjustments to accommodate those changes

10 CHAPTER TEN

10.1 Case Study

In order to study the impact of the Six Sigma program, a study was conducted in the Printed Circuit Assembly operation. The assembly line for printed circuit board is a combination of machine and manual processes. The machines place solder paste on the board and then places component as required. The board is run through an oven to melt the solder and make a permanent connection between the component and the surface of the board. In this study two identical processes are compared, one operates at Three Sigma level and the other operates at Six Sigma level. Data are collected over a month and a half period of time, three runs a day - fifteen runs a week, totaling ninety runs. Out of these ninety samples, forty five samples were chosen randomly. The type of research conducted was quantified, the data derived through a series of design of experiments and the optimum points were evaluated.

Simply the printed circuit assembly manufacturing requires all electronic components to be interconnected to the printed circuit boards. The connection process is the main challenge for the PCBA industry. Basically two major processes have been evaluated, : first, surface mount and second, through hole. In surface mount the electrical components are soldered to pads on the surface of the printed circuit board. In through hole the components are soldered in the plated hole drilled through the boards. These connections are called the solder joints in this research. The quality of the solder joints are evaluated using IPC 610-B standard. The samples are taken from three categories of boards. The complexity factor will distinguish between the samples.

The categories of boards are:

1.   Simple boards (a) 1 - 500 solder joints

2.   Average boards (b) 500 - 1000 solder joints

3.   Complex boards (c) 1000+ solder joints

10.2 Sampling plan

A Stratified Random Sampling is chosen. Three subpopulations of the same size are chosen. The first sample size (SS) was thirty units for each categories. Figure 1. shows the overall sampling plan for these three categories for the Three Sigma (3S) experiment. A similar table will be developed for the Six Sigma sampling plan.

Figure 1. Stratified Sampling Plan for Three Sigma

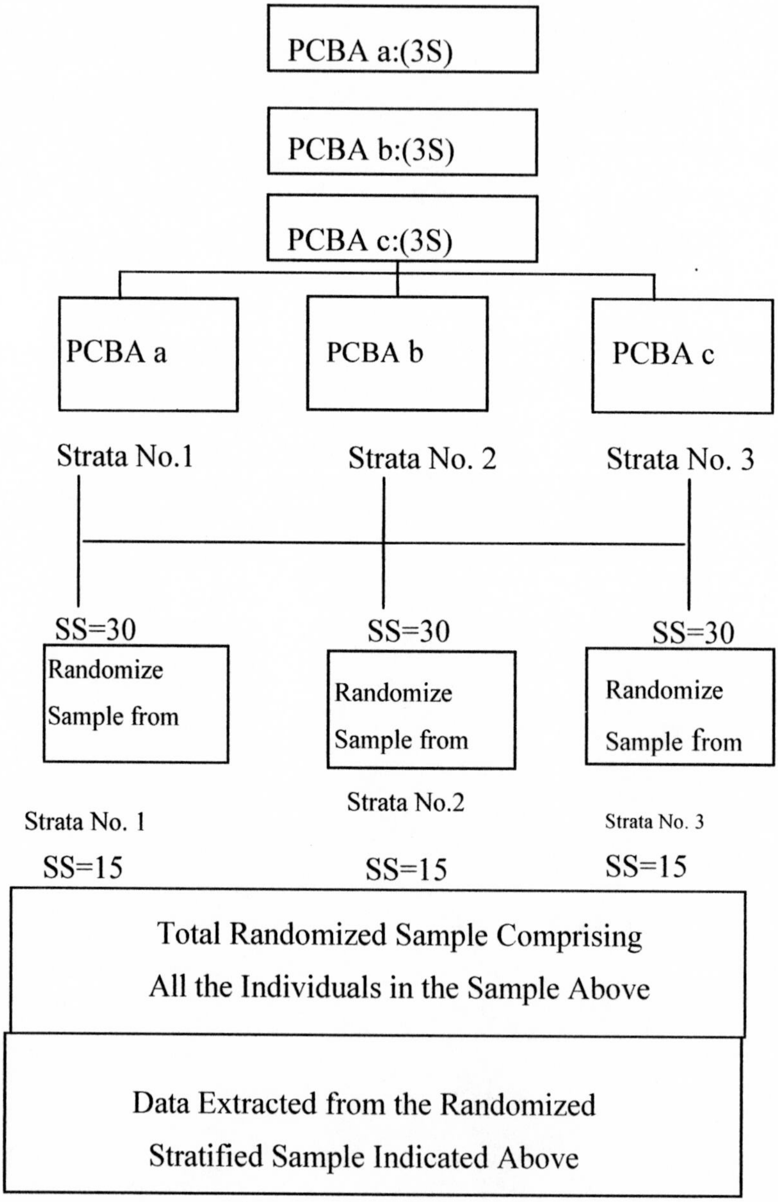

PCBA a:(3S)

PCBA b:(3S)

PCBA c:(3S)

PCBA a	PCBA b	PCBA c

Strata No.1   Strata No. 2   Strata No. 3

SS=30   SS=30   SS=30

Randomize Sample from	Randomize Sample from	Randomize Sample from

Strata No.2

Strata No. 1   Strata No. 3

SS=15   SS=15   SS=15

Total Randomized Sample Comprising
All the Individuals in the Sample Above

Data Extracted from the Randomized
Stratified Sample Indicated Above

The three different categories have been run with the existing process that operates at the Three Sigma level. The respond as mentioned previously, is Defect Per Million Opportunity (DPMO). The "lower the better" will determine the performance of the process.

After completing the production of the boards, fifteen samples of each categories are taken and the DPMO level is measured. Table 1 shows the DPMO for each category for fifteen samples. The PCBs have been inspected by the same inspectors Each board is checked for solder bridge, insufficient solder, missing component, reversed component, wrong component, cold solder, open solder, damaged components and any other workmanship problems which have been specified in the Industry Standard Handbooks (IPC 610-B).

10.3 *Measurement Techniques*

In order to compare these two processes, the measurement method needs to be defined. The indicator will be defect per million opportunity (DPMO) which considers the complexity factor for each PCB. DPMO is calculated as follow:

$$DPU \text{ (Defect Per Unit)} = \frac{\# \text{ of Defect}}{\# \text{ of PCB}}$$

$$Multiplier = \frac{1,000,000}{\# \text{ 0f opportunities}}$$

DPMO = Multiplier * DPU

After calculating the DPMO for each PCBA, the "lower is the better" will determine the better result and the better process. For each category and each sample the same formula is used. Based on the sterified sampling plan fifteen samples have been chosen. Table 1 shows the DPMO for each sample in the three categories.

Table 1

Three Sigma DPMO

Sample	1	2	3	4	5	6	7	8	9	10	11	12	13	14	15
Simple	2500	2650	2760	2450	2630	2211	2811	2740	2750	2550	2320	2670	2690	2860	2590
Average	1850	1870	1790	1758	1840	1855	1900	1868	1795	1780	1790	1820	1832	1839	1798
Complex	320	310	310	305	322	315	298	280	250	290	295	311	290	282	279

For example the DPMO for the first sample in the "simple" category is 2500 and for the fifth sample in the "complex" category are 322.

In order to have a better understanding of the numbers the standard deviation and mean are calculated to answer the following questions:

1.   How far are we from the target?

2.   What is the quality level?

3.   What is the associated cost with the variations?

At this point it was decided to design a series of experiments to modify and evaluate the existing factors, modify the main effects and design a process to the Six Sigma standard. Table 2 shows a significant difference between the target and outgoing quality level for the Three Sigma process.

Comparing these three categories indicates that the more complex boards have fewer variations.

Table 2

Calculation for mean and standard deviation

Category	Mean	Target	Std
Simple	2612.13	2700	181.03
Average	1825.66	2700	39.73
Complex	297.13	2700	19.32

10.4 Process Modification

In control process is capable of preventing some of the problems. The major contributors of these discrepancies need to be identified and eliminated. For this reason, a series of experiments were designed and conducted to evaluate the Three Sigma process problems.

In order to evaluate and modify the existing process and comply with the characteristic elements of the process capability ratio, the Taguchi method for the design of experiment is chosen. The Taguchi method is a mix of several tools that have been developed to optimize product and process performance. It combines elements of brainstorming, designed experiments, orthogonal arrays, analysis of variance and Dr. Taguchi's new term "loss function".

10.4.1   Orthogonal Arrays

Orthogonal arrays have several unique qualities. They can be at several levels, but the most popular are either two or three levels. One of the most important attributes is that they are balanced. While the level is constant in a column, all of the other levels in the other column will be rotated through all levels.

In order to evaluate the main effect for the variables, L8, which is a two level array, has been chosen L8 = 27 =128. N factors at two levels would normally require 2 to the power N experiments. Using orthogonal array maximizing experiments can be accomplished with N+1 experiments. For example, in the L8 array example, all experiments for 7 factors can be done in eight experiments

versus 2 to power 7 or 128 experiments in the traditional system of varying one factor at a time.

Objective. The two main processes for the printed circuit assembly (Surface Mount Technology (SMT) and Wave Solder) are evaluated. The boards that have been chosen run through these two processes. First the Design of Experiment (DOE) is conducted for the SMT process and then for the Wave Solder.

10.4.2   Design of Experiments

The existing process is still undergoing major improvements. Design variability has a big effect on yield and is being studied by a joint team including the supplier. Yields during the last 30 batches have ranged from 75% to 95%, with big swings with each new feedstock shipment.

Experimental variables

Response variables	MeasurementTechnique
Defects per Million Opportunity (DPMO)	Visual Inspection

Factors under study	Levels
1. Machine Parameters	1,2,3

10.4.2.1 *Replication:*

There is enough supply of the experimental boards to make sample from each. The study should be completed within a month. Six replications (blocks) of each of the three boards are processed.

10.4.2.2 *Method of randomization:*

Table of random permutations of 4 was used to assign the three PCBA to the four batches made from each process..

10.4.2.3 *Data collection forms:*

Data collection sheets. (attach copy)

10.4.2.4 *Planned methods of statistical analysis:*

Run charts of DPMO; run chart of DPMO adjusted for background variables.

10.4.2.5 *Estimated cost, schedule, and other resource considerations:*

Initial study can be completed during one month of operation; possible losses or gains in DPMO from the different categories of PCBA.

10.4.2.6 *Design of Experiment for the SMT process*

To optimize manufacturing capability for these two experiments, an L9 orthogonal array will be utilized in this evaluation, targeting a "smallest is best" output.

The output measure will be paste height after printing. The objective is not only to consistently print the target value at 0.005″, but also to minimize the standard deviation.

The process used for this experiment will be the FUGI "GSPII" stencil printer, and three different test PCBs. Table 3 describes the required material and equipment for the experiment.

Table 3

The required equipment list

Solder Paste	Alpha WS609
Stencil Printer	FUGI "GSPII"
Stencil Manufacturer	0.005″ thick
Test boards	PCBAa, PCBAb, PCBAc
Depth Gauge	Cyberoptics Microscan

Table 4 illustrates different factors for the screen printing process. The experiment is designed to determine the best stencil printing machine set ups for use in the stencil printing process. In this experiment there are four controllable factors of interest, each with three levels.

Table 4

The Stencil Printing Matrix

			Levels		
Letter Factor		Description	(1)	(2)	(3)
(A)	Squeegee Speed	The speed the squeegee move during the print cycle in inches per second.	.5	1.5	2.5
(B)	Squeegee Downstop	The distance the squeegee would travel beyond the top of the PWB if not restrained by the PWB in inches.	.030	.060	.080
(C)	Snap-Off Distance	The distance from the top of the substrate to the bottom of the stencil prior to printing in inches.	.010	.020	.030
(D)	Squeegee Pressure	The pressure which is applied to the squeegee during printing in pounds.	30	45	60

Two additional factors shall be held constant during the experiment. These constants are: Stencil Cleanliness for which the top, bottom and side surfaces of the stencil should not have contamination of paste and for which the Vacuum shall be on during printing.

Table 5

Orthogonal Array (L9) for Stencil Printing

Std. Order	Factor A B C D	Squeegee Speed (in/s)	Down Stop (in)	Snap-Off Distance (in)	Squeegee Pressure
1	1 1 1 1	0.5	.03	.01	30
2	1 2 2 2	0.5	.06	.02	45
3	1 3 3 3	0.5	.08	.03	60
4	2 1 2 3	1.5	.03	.02	60
5	2 2 3 1	1.5	.06	.03	30
6	2 3 1 2	1.5	.08	.01	45
7	3 1 3 2	2.5	.03	.03	45
8	3 2 1 3	2.5	.06	.01	60
9	3 3 2 1	2.5	.08	.02	30

The measurement of the response output shall be visual printing, alignment of the solder paste on the substrate, and measurement of printed paste height. This response is to determine optimal mixed technology (fine pitch, standard pitch and through hole component) for the machine parameters using the FUGI Stencil Printer. The decision to use this machine based on the capability to print fine pitch patterns in a production mode. The following factors must be kept as constant as possible:

A.   Solder Paste. Open jar and thoroughly stir paste before use. Paste must have a smooth consistency.

B.   Stencils. Inspect stencils to ensure they meet the specifications.

180

C.   Squeegee. Visually inspect the squeegee to ensure the edge is good the blade is properly adjusted in the holder. The squeegee must provide good wiping of the stencil.

D.   Test Boards. Number samples.

E.   Machine Calibration. Ensure the squeegee head, stencil holder and the print table is parallel to each other.

Thirty samples shall be processed for each set of experimental run. Additional replication may be run if the results of the experiment are questionable.

Stencil is wiped, cleaned between prints, the thickness is measured at four places per PWB and the data is recorded.

Acceptability Testing. As data shall be evaluated at the end of this experiment, the best machine parameters shall be selected and utilized in production.

10.4.3   Design of Experiment for the Wave Solder Process

Parameter Interaction Selection. In selecting the parameter interactions, a special table was constructed to look at all possible pair wise interactions to select the ones judged most likely to affect the process and therefore to be included in the analysis. Additionally, the interaction discussions facilitated the assignment of the three parameters that could not be assigned to primary columns. The interaction table generated was triangular and similar to mileage chart tables on geographical maps. Twelve pair wise interactions were selected for further analysis (Table 6).

Interaction was the most confusing element of the Taguchi methodology. In subsequent Taguchi experiments, I have favored the use of no interacting orthogonal array tables such as L12, L18, and L36.

Table 6: Interaction Tables for wave solder process Taguchi Experiments Pair wise Interactions Columns of L32 are denoted by ( )

	Conv Angel (2)	Wave Temp. (4)	Direction (6)	Wave Height (8)	Preheat Temp (16)	Wave Width (18)	Conv Speed (30)
Flux (1)	(3)	(5)	0	(9)	(17)	0	0
Conv Angle (2)		0	0	(10)	0	0	(28)
Wave Temp (4)			0	(12)	(20)	0	(26)
Direction (6)				0	0	0	(24)
WaveHeight (8)					0	0	(22)
PreheatTemp((						0	(14)
Wavewidth (18)							0

It was decided to carry on the experiment for a short term, to minimize the effect of environmental conditions on the outcome of the results and to reduce the production schedule interruption. Several modifications were made to the equipment in order to change the parameters quickly. Important to the success of the experiment was the clear identification of each PCB, and the extensive training of the operators and supervisors and to gain their support. The experiments were performed in two 8-hour shifts.

Data Analysis. The L32 array analysis using all 8 parameters was done on the average defects per experiment and on the signal to noise ratio as defined by "smaller is better." A quick check of the results shows the dramatic effects obtained by varying the parameters. Solder Bridges ranged from an average of 0.25 short per board to 26.70 and Excessive Solder averaged from 0.00 to 45.30 per PCB.

Rearranging the data by parameter number, two tables (Tables 7 and 8) of the average solder bridge and excessive solder were generated using the two levels of each parameter. Parameter contribution is the effect of selections the correct parameter level on the total parameter average (+ for level 1, - for level 2).

Table 7 Average of All Parameters, Solder Bridge/Test PCB

Parameter	Level 1 Avg. Solder Bridge	Level 2 Avg. Solder Bridge	Parameter Contrib.
Direction	13.14	4.62	-4.26
Flux	5.30	12.46	+3.58
Speed	7.89	9.87	+0.99
Wave temperature	9.55	8.21	-0.67
Wave height	9.25	8.51	-0.37
Conv. angle	8.54	9.22	+0.34
Preheat temp.	9.11	8.65	-0.23
Wave width	9.09	8.67	-0.21

Table 8

Average of All Parameters, Excess Solder/Test PCB

Parameter	Level 1 AV. Excess Solder	Level 2 AV. Excess Solder	Parameter Contribution.
Flux	12.48	29.72	+8.62
Direction	23.60	18.16	-2.50
Height	23.57	18.63	+2.47
Angle	23.23	18.97	-2.13
Wave temperature	19.88	22.32	+1.22
Speed	20.30	21.90	+0.71
Wave width	20.39	21.81	+.071
Preheat temp.	20.84	21.36	+.026

ANOVA analysis was performed. The analysis included calculating the F factors and % contributions for significant parameters. All analyses were done on bridge and excessive solder independently (Table 9,10). The percent contribution is another of Dr. Taguchi terms. It is calculated by subtracting the error variance from each parameter's mean square, then normalizing it to 100%.

Table 9 ANOVA Table for Mean Solder Bridge Defects

Parameter	DF	Sum Of Square	Mean Square	F Ratio	Percent Contrib.
Flux	1	410.34	410.34	29.65	29.80%
Angle	1	3.72	3.72	0.27	
Wave Temp.	1	14.32	14.32	1.03	
Direction	1	581.49	581.49	42.02	44.66%
Wave height	1	4.39	4.39	0.32	
Preheat temp.	1	1.72	1.72	0.13	
Wave width	1	1.40	1.40	0.10	
Conv. speed	1	31.11	31.11	2.25	
1 X 2	1	3.39	1.39	0.24	
1 X 3	1	0.30	0.30	0.02	
1 X 5	1	11.70	11.70	0.85	
1 X 6	1	3.77	3.77	0.27	
Sum Total	12	10678.68	88.97		

Pooled erro	12	262.93	F Table = 4.38 @ 95%	
Total	31	1330.61		

Table 10

ANOVA Table for Mean Excess Solder defects

Parameter	DF	Sum Of Square	Mean Square	F Ratio	Percent Contribute
Flux	1	2377.05	23377.50	77.53	55.70%
Angle	1	145.35	145.35	4.74	
Wave Temp.	1	47.78	47.78	1.56	
Direction	1	200.50	200.50	6.54	4.03%
Wave height	1	195.23	195.23	6.37	3.91%
Preheattemp	1	2.21	2.241	0.07	
Wave width	1	15.96	15.96	0.52	
Conv. speed	1	20.58	20.58	0.67	
X 2	1	390.60	390.60	12.74	
X 3	1	54.86	54.86	1.79	
X 5	1	175.97	175.97	5.74	3.45%
X 6	1	4.20	4.20	0.14	
Sum Total	12	3630.29	30.66		
Pooled error	12	582.59		F Table = 4.38 @	

Figure 2. Taguchi analysis for wave solders process mean parameter effect. Level 1, level 2 By pooling the smaller errors, the Excess and Bridge defects are averaged.

"Table 11" ANOVA calculations would be shortened as shown in Table 10

Parameter	DF	Sum Of Squares	Mean Square	F Ratio	Contribution	Confidence Factor%
Flux Type	1	410.34	410.34	35.13	29.96%	99%
Direction	1	581.49	581.49	49.79	42.82	99
Sum Errors	29	338.78	11.86			
Total Errors	31	1330.61				
Flux Type	1	2377.05	2377.05	81.61	55.73	99
Conv. Angle	1	145.35	145.35	4.99	2.76	90
Direction	1	200.50	200.50	6.88	4.07	90
Wave Height	1	195.23	195.23	6.70	3.94	90
Flux x Ang.	1	390.60	390.60	13.41	8.58	99
Flux x Hgt.	1	175.97	175.97	6.04	3.49	90
Error Sum	25	728.18	29.13			
Total Error	31	4212.88				

10.4.4   Setting Parameter Levels

In selecting the final parameters for the production process, the analysis had to take into account the significant factors derived from the mean and signal to noise calculations. Several additional parameters, which were previously thought to be important, were not significant either in mean or S/N analysis. Since the flux was the dominant factor, half of the L32 was taken and analyzed as an L16 array, to verify more significant factors. For the two parameters that were not significant in either L16 or L32 analysis, it was decided to use the L16 half of the data to set these two parameters. Preheat and wave width was not significant in either case, and they were arbitrarily set.

Table 12ANOVA Table for S/N Excess Solder Defects

	Parameter	DF	Sum Of Squares	Mean Square	F Ration	Percent Contribution
1	Flux	1	78.79	78.79	2.75	
2	Angle	1	2.69	2.69	0.31	
3	Wave temp	1	0.86	0.86	0.01	
4	Direction	1	44.26	44.26	1.53	
5	Wave height	1	95.83	95.83	3.34	
6	Preheat temp.	1	17.13	17.13	1.98	
7	Wave width	1	3.59	3.59	0.41	
8	Conv. speed	1	169.85	169.85	5.92	11.93
9	1 x 2	1	135.66	135.66	4.73	9.04
10	1 x 3	1	3.20	3.20	0.37	
11	1 x 5	1	15.74	15.74	1.82	
12	1 x 6	1	71.50	71.50	2.49	3.62
	Error Total	19	544.70	28.67		
	Sum Total	31	1183.810			

There were not significant factors for S/N Excess Solder Defects.

Table 13:Setting parameter level to optimum condition

Parameter	L32 Mean L16				L32 S/N L16				Final Level
	SB	ES	SB	ES	SB	ES	SB	ES	
Flux	1	1							1
Conv. angle		2		2			1	2	2
Wave temp.			2	1			1	2	2
Direction	2	2	2				2	2	2
Height		1					1		1
Preheat									2
Wave width									1
Conv. speed			1		1		1		1

By combining all information from the ANOVA analysis of experiments, each parameter level could be set to optimum conditions

(Table 12). Most of the data analyses of the experiments were consistent in determining the optimal level for each parameter. In case of different levels, the choice was made to lower the mean (smaller is better) of the excessive solder defects since they were more common on production boards. The preheat and the wave width parameters were not significant, but were assigned their values based on the interaction plots.

10.5 *The Six Sigma Approach*

Based on the statistical data provided by the Taguchi experiment, the predicted levels of defects using the test PCBs were calculated at an average of one excessive solder from the same lot as the experiment boards, using the recommended parameters. The confirming experiment yielded zero excessive solder and two solder bridge.

When other confirming experiments were run using test boards assembled at different times, the defect rates varied, indicating the effect of other control variables such as lead angle and lead length that had not been considered in the experiment or the effect of other interactions not studied. Feedback was given to the assembly department on the effect of the variations of their processes.

After the completion of the design of experiments, some of the parameters in the processes have been modified in order to comply with the Six Sigma approach. The method to approach the Six Sigma concept is accomplished through the following elements:

1.   Product characteristics
2.   Product elements
3.   Process steps
4.   Nominal design values
5.   Determining the process capability
6.   If Cp is not ≥ 2 or (Cpk = 1.5) change the process or product design

$$Cp = \frac{\text{Design specification width}}{\text{Process width}} \qquad Cpk = Cp(1-k)$$

$$k = \frac{\text{Target - Actual}}{1/2(\text{USL - LSL})}$$

193

Product design has a goal of increasing tolerance to the level which will still permit successful functioning of the product. Process design has the goal of minimizing the variability of the process which reproduces the characteristics required for successful functioning of the product and for keeping the process on the target (nominal) value of the characteristic. Figure 3 illustrates the stratified sampling plan for the Six Sigma process.

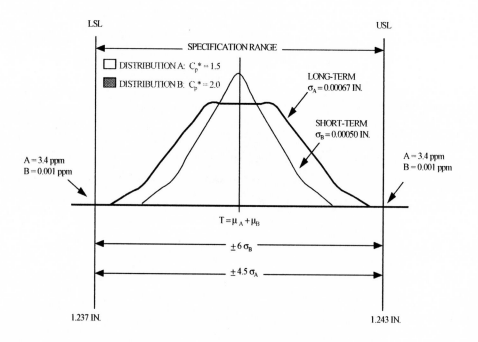

Figure 3. Comparison of two different distribution

It is often important to determine how well a process performs relative to a specification.

Process capability, the measurement for this assessment, is often reported using a Process Capability Index (Cp) and Process Capability Index (Cpk).

Motorola uses Process Capability Index (Cp) and Process Capability Index (Cpk) within their Six Sigma program. An internal or external process is considered to be Six Sigma if Process Capability Index (Cp) is equal to or greater than 2.0 and Process Capability Index (Cpk) is equal to or greater than 1.5. A process with these indices has a calculated defect rate of 3.4 parts per million.

When using process indices to describe the quality of a process, we need keep in mind that this calculation quantifies the tails of a distribution and the uncertainty of the result from this calculation is typically very large. To illustrate how this can impact an organization, consider using reported process capability number(s) to choose the best supplier. Differences between reported numbers are not all attributable to process quality. Issues often not considered when comparing reported process indices is sample size, data normality, data outliers, and samples not being representative of future product. One supplier might report a very good process capability value using only five samples produced on one day, while another supplier of the same commodity might report a somewhat lesser process capability number using data from a longer period of time that more closely represents the process. If we were to only compare these two process index numbers when choosing a supplier, the best supplier might not be chosen.

It is often more meaningful when comparing and quantifying process capability to rely on a slightly different approach. With this approach, there is no need for specifications hence, management can use this approach to describe their important business processes. In addition, this measurement approach can be an integral part of a process quantification and improvement program.

This process measurement approach consists of two steps. First, a control chart identifies special cause problems for resolution. Secondly, a normal or Weibull probability plot of the common cause variability from the control chart describes the capability of the process as a picture.

Consider how difficult it is to physically picture the capability of a process that, for example, has reported Process Capability Index (Cp) = 1.6 and Process

Capability Index (Cpk) = 1. A probability plot overcomes this confusion by describing the process in terms that are easy to understand. With this tool, you can make definitive statements and pictorially describe the capability of a process. Specifications and normality are not required. Also, this analyses can detect outliers, address bimodal distributions, and consider more than one distribution for a process.

It often appears that organizations get so involved in trying to determine the capability of their process that they loose sight of an important issue — identifying and implementing improvement opportunities. Measurements are useful to help baseline current processes however, measurements alone don't improve anything. "Meaningful Measurements & Effective Process Improvements" describes a roadmap for integrating measurements with process improvement activities. The referenced articles can give insight to efficient approaches for measurements and improving customer satisfaction.

Figure 3 illustrate the area under normal; curve distribution for two different process

The limitation of the experiment did not allow us to have any control on the first two elements (were Product Characteristics and Product Elements). It means the product characteristic and the product elements are not a variable in this experiment. We use the same product with the same design and elements for both processes. After conducting all the design of experiments, the mean effects have been evaluated and have been reduced. Table 15 shows the DPMO result from of the Six Sigma after process modification. The same experiment was conducted with taking fifteen samples and comparing the DPMO for each one.

Table 14

DPMO result from the Six Sigma process

Sample	1	2	3	4	5	6	7	8	9	10	11	12	13	14	15
Simple	218	262	252	240	250	200	250	265	211	282	272	250	231	222	241
Average	110	111	114	125	115	110	117	127	129	131	120	124	115	116	115
Complex	50	52	50	50	51	52	51	49	49	51	52	50	52	52	52

Table 15

Calculation of mean and standard divination for Six Sigma process

Boards	Mean	Target	Std
Simple	243.06	3.4	23.19
Average	118.6	3.4	6.96
Complex	47.4	3.4	1.12

Table 15 shows the results from the respond factors. As it shows the mean and variation have been reduced compared with the similar table for the Three Sigma dramatically.

10.6 Result and Conclusion

Data Analysis As was discussed in chapter 3, a series of experiments were designed and conducted to evaluate the existing normal process and the main contributing factor for each of the manufacturing steps. The first and one of the most important factor was the screen printing process which deposits the solder paste on the boards. Four factors are chosen and the results are shown in the following tables.

Table 16. Evaluates all the factors in the Screen Printing Process.

Table 16:

Design of Experiment result from the Stencil printing process

Std Order	Factor A B C D	Squeegee Speed (in/s)	Down Stop (in)	Snap-off Distance (in)	Squeegee Pressure (#)	#1	#2	AVG
1	1 1 1 1	0.5	.03	.01	30	7.0	7.4	7.2
2	1 2 2 2	0.5	.06	.02	45	5.5	5.7	5.6
3	1 3 3 3	0.5	.08	.03	60	4.4	4.2	4.3
4	2 1 2 3	1.5	.03	.02	60	5.2	5.8	5.5
5	2 2 3 1	4.5	.06	.03	30	5.5	5.7	5.6
6	2 3 1 2	1.5	.08	.01	45	5.8	5.6	5.7
7	3 1 3 2	2.5	.03	.03	45	6.8	6.6	6.7
8	3 2 1 3	2.5	.06	.01	60	5.2	5.2	5.2
9	3 3 2 1	2.5	.08	.02	30	5.6	5.4	5.5

$C.F = (Sum\ y)^2 / n = (7.2 + 5.6 + 4.3 + 5.5 + 5.6 + 5.7 + 6.7 + 5.2 + 5.5)^2/g = 302.75$

$S_{total} = (7.2^2 + 5.6^2 + 5.2^2 + 5.5^2 + 5.6^2 + 5.7^2 + 6.7^2 + 5.2^2 + 5.5^2)^2/g = 3.74$

$$S_A = \frac{(7.2 + 5.6 + 5.2)^2}{3} + \frac{(5.5 + 5.6 + 5.7)^2}{3} + \frac{(6.7 + 5.2 + 5.5)^2}{3} - 302.76 = .24$$

$$S_B = \frac{(7.2 + 5.5 + 6.7)^2}{3} + \frac{(5.6 + 5.6 + 5.2)^2}{3} + \frac{(5.2 + 5.7 + 5.5)^2}{3} - 302.76 = 1.99$$

$$S_C = \frac{(7.2 + 5.7 + 5.2)^2}{3} + \frac{(5.6 + 5.5 + 5.5)^2}{3} + \frac{(5.2 + 5.6 + 6.7)^2}{3} - 302.76 = .37$$

$$S_D = \frac{(7.2 + 5.6 + 5.5)^2}{3} + \frac{(5.6 + 5.7 + 6.7)^2}{3} + \frac{(5.2 + 5.5 + 5.2)^2}{3} - 302.76 = 1.14$$

Table 17

ANOVA for the four factors

	DF	^{S}T	V	F	P%
A error	2	.24	.12		
B	2	1.99	.995	8.29	
C	2	.37	.185	1.54	
D	2	1.14	.57	4.75	

Table 18

Comparing the four factors

	A	B	C	D
Level 1	6.0	6.5	6.0	6.1
Level 2	5.6	5.45	5.53	6.0
Level 3	5.8	5.45	5.87	5.3
	17.4	17.4	17.4	17.4

17.4 / 3 = 5.8

Resultant Effect Predictions: B&D Selected from Pooled Table.

Average = 5.8

Average of Significant Factors:

$B_1 = 6.47$ $B_2 = 5.47$ $B_3 = 5.47$

$D_1 = 6.10$ $D_2 = 6.00$ $D_3 = 5.30$

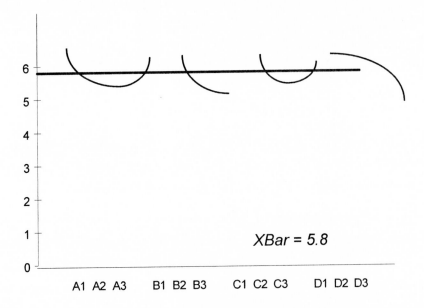

Figure 5. Trend diagram of all factors

This design of experiment (DOE) intended to evaluate and optimize manufacturing capability for fine pitch stencil printing of solder paste using three different categories of boards (simple, average and complex). The experiment was successful in that the output of the optimal parameter settings was close to the target which will result in high quality solder joints. The DOE established optimal machine parameters using the FUGI-GSPII stencil printing machine without two dimensional vision. The optimum levels for squeegee speed, downstop, and snap-off (factors A, B, and C) were at level two. The squeegee pressure (Factor D) was level three. The ANOVA analysis shows that the significant factors were squeegee downstop (B), and squeegee pressure (D) and that these two factors influenced as much as 75% of the paste printability. The remaining two factors should be pooled into the error. Our hypothesis is that the 25% not accounted for in the ANOVA analysis, is the result of interaction between factor (B) downstop, and factor (D) squeegee pressure and will be evaluated as such. The optimum parameter setting will be used to design and conduct the Six Sigma process.

10.6.1   Results of the Design of Experiment for Wave Solder Process

The wave solder parameters were set according to the analysis of the Taguchi experiment. Daily inspection of 100% of production of PCBs was performed and the results averaged as solder defects in parts per million.

This Taguchi experiment provided a very efficient method for achieving a high-quality wave soldering process with minimum investment in engineering time and resources. The reduction of soldering defects at the factory led to a confirmed production savings of $75,000 per year, based on the results of the first 6 months. The initial investment was $10,000 for a run of 180 assembled test PCBs.

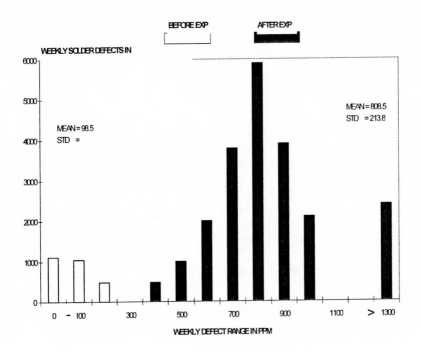

Figure 6. Quality level at wave solders process in production one-month histogram.

	Mean	SD
Before	808.50	213.80
After	98.50	55.30

The savings implications for the Electronics Industry in the United States could be tremendous based on the amount of quality improvement we observed. The statistical estimated an annual volume of 265 billion solder connections in 1986, or $100 million dollars annually in savings of rework, inspection, material and solder metal.

205

Lessons learned in this successful implementation were to choose a fairly difficult and persistent problem, to select a wide ranging and experienced process team, to obtain management support early in the book, and to minimize the importance of interactions in the design of the experiment.

10.7 Quality Loss Function

The quality loss function was calculated based on the formula by Dr. Genichi Taguchi, for smaller is better. Figure 6 illustrates the graphical view of the Quality loss Function. The formula is defined as follow:

$L(x) = k(\delta x + (\mu - m)^2$

$L(x)$ = Expected loss per piece of quantity x

K = Loss coefficient

δx = Standard deviation of the process

μ = The actual process mean

m= The target

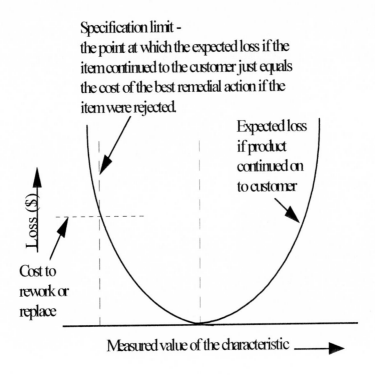

Specification limit -
the point at which the expected loss if the
item continued to the customer just equals
the cost of the best remedial action if the
item were rejected.

Figure 7. Taguchi Quality Loss Function Curve

The cost of locating and removing a solder bridge or an excessive is estimated at $0.06 per defect, based on the inspection and test system analysis. Prior to the experiment we had the following means and standard deviation for the Three Sigma process.

Table 19

Calculation for mean and standard deviation for the Three Sigma process

Category	Mean	Target	Std
Simple	2612.13	2700	181.0
Average	1825.66	2700	39.73
Complex	297.13	2700	19.32

Based on the mentioned numbers for the Three Sigma process the quality loss function is applied to analyze the cost of the process.

By applying the quality loss function in Figure 6 the cost for different categories are calculated as follows:

Simple boards: $L(x) = 0.06(181.0355+(2612.133-2700)^2) = \$47.41$

Average boards: $L(x) = 0.06(39.73603+(1825.667-2700)^2) = \$458.65$

Complex Boards: $L(x) = 0.06(19.32011+(297.133-2700)^2) = \$3,462.82$

The same approach is used for the Six Sigma process. Table 20 shows the standard deviation, target and the mean for three different categories. The natural variation of the process defines the limits within which the vast majority of the measurements will occur.

By convention, the natural variation of a performance characteristic is equal to twelve times its short term standard deviation and, therefore, includes 99.999998 percent of the items in the population if its distribution is normal.

Table 20

Calculation of mean and standard divination for the Six Sigma process

Boards	Mean	Target	Std
Simple	243.06	3.4	23.19
Average	118.6	3.4	6.96
Complex	47.4	3.4	1.12

The following calculation is the result of Taguchi loss quality function for the Six Sigma process. Once again the same cost structures require at $0.006 to fix any non conformance product.

Simple boards: $0.06(23.19318+(243.0667-3.4)^2) = \$344.58$

Average boards: $0.06(6.96727+(118.6-3.4)^2) = \$79.66$

Complex boards: $0.06(1.121714+(47.4-3.4)^2) = \$11.62$

The cost is usually a driving factor. The Taguchi Loss Function is expressing the result clearly. For the simple boards the cost of quality is higher with Six Sigma than three sigma. But for the average and complex boards the Six Sigma shows a significant difference from Three Sigma to Six Sigma.

10.8 Conclusion

From a conclusive analysis it appears that the road map to 6 sigma is a simple one, the difficulty is in the execution. The major difference is cost saving and the outgoing quality level.

In reference to Tables 20 and 21, the standard deviation is the driving factor. Less variation results in high quality with less cost. In order to have a minimum variation all process stages must be shown to be in statistical control, then the variations of the process need to be identified and means obtained to reduce or eliminate them. This can be done by determining the critical parameters of the process through experience or by statistical design of experiment, identifying the source of variation that effect and either reducing or eliminating them.

Six Sigma also considers all the process parameters which could caused any variation to the process, sensitivity of this scale allows us to quantify the variations and reduce them. Any process which is running and fitting within 6 sigma could guarantee minimum defects at less then 3.4 per million.

Finally for a product to be built virtually defect-free, it must be designed to accept characteristics which are significantly more than ±3 sigma from the mean. A design which can accept twice the normal variation of the process, or ± 6 sigma, can be expected to have no more than 3.4 defective parts per million, even if the process mean were to shift by as much as ±1.5 sigma.

The study concludes that the printed circuit assemblies in the "Simple" category are not cost effective to be produced in the Six Sigma process. The "Complex" and "Average" categories are the best candidates for the Six Sigma process. The experience at this industry indicates that the simple boards are easy to troubleshot and fix. From visual inspection point of view it's very simple to inspect less than 500 solder joints at any given process.

References

Allderge, M., (1993). The Six Sigma Concept. New York: McGraw-Hill.

American Society for Quality Control. (1991). Statistics Issues. New York: Macmillan Circuit Assembly magazine. (1995). Electronics Manufacturing Industry. San Francisco, CA.

Campbell, D. T., Stanley, J. C. (1963). Experimental and quasi-experimental design for research. Houghton Mifflin Company.

Daniel, C. (1976). Application of statistics to industrial experimentation. New York: John Wiley and Sons.

Dobyns, L. & Crawford, C., (1991). Quality or else. Boston, MA: Houghton Mifflin Company.

Gill, M., (1990). Staking Six Sigma. Tucson, AZ: Motorola University.

Hald, A. (1952). Statistical theory with engineering applications. New York: John Wiley and Sons.

Harry, M., & Lawson, R.., (1992). Six Sigma producibility analysis and process characterization. Reading, MA: Addison-Wesley.

Hogg, R., & Gray, A. V. (1989). Introduction to mathematical statistics. New York: McGraw-Hill.

Juran, J. M., (1992). Quality by design. New York: A Division of Macmillan Inc.

Koon, G., (1993). The Six Sigma Approach. Tucson, AZ. Motorola University.

Johnson, L.G. (1972), Theory and techniques of variation research. New York: Elsevier Publishing Co.

Lipson, C., & N. Sheth, (1973). Statistical design and analysis of engineering experiments. New York: McGraw Hill Book Co.

La Brake, M. L. (1992). Test for differences. Tucson, AZ: Motorola University.

Moura, E., (1991). How to estimate failure rating in life testing. Milwaukee, WI: BookCraft.

Motorola Semiconductor Products Sector. (1992). Reliability & quality, Reading, MA: Addison-Wesley.

Motorola Cooperate Office. (1987). What is Six Sigma? Tucson, AZ: Motorola University.

Motorola University Press. (1993). Six Sigma process by Design. Tucson, AZ: Motorola University

Nelson, (1988). Control charts: Relation subgroups and effective applications. New York: Mack Printing Co.

O'Connor, P., (1991). Practical reliability engineering. New York: John Wiley & Sons, Inc.

Phadke, M., (1989). Quality Engineering Using Robust Design. Englewood cliffs, NJ: Prentice-Hall.

Research Institute of Motorola. (1993). Basic Six Sigma concept. Tucson, AZ: Motorola University Author

Ross, P., (1987). Taguchi Techniques for Quality Engineering. New York: McGraw Hill.

Standard IPC-PC-90. (1989). General requirements for implementation of statisticalprocess control. Illinois: Institute for Interconnecting and Packaging Electronic Circuits.

Taguchi, G., (1988). Introduction to Quality Engineering, Designing Quality into Products and Process. Asian Productivity Organization.

Taguchi, G., (1988). El Sayed, and Hsiang, Quality Engineering in Production Systems. New York: McGraw Hill.

Tobin, M., (1993). <u>Statistical Process Control</u>. New York: McGraw Hill.

Therrien, G., (1989). <u>Process and Productivity</u>. Reading, MA: Addison-Wesley.

Walpole, M., (1989). <u>Probability and statistics</u>. New York: Macmillan. U.S.A. Today. (1989). <u>Motorola Approach to Six Sigma</u>. New York.

Wade, O. R., (1967) <u>Tolerance control in design and manufacturing</u>. New York: Industrial Press, Inc.

Wick, C. & Raymond, F. (1987), <u>Tool and manufacturing engineers handbook</u>. Dearborn, Michigan: McGraw Hill Book Co.

Wu, Y., & Moore, W.H. (1985). <u>Quality Engineering, Product and Process Design Optimization</u>, New York, American Supplier Institute Inc.

F. W. Breyfogle III, Statistical Methods for Testing, Development, and Manufacturing, Wey, New York, NY, 1992.

F. W. Breyfogle III, Implementing Six Sigma: Smarter Solutions using Statistical Methods, Wiley, New York, NY, 1999.

F. W. Breyfogle III, "Implementing Six Sigma," ASQ Quality Management Forum, 25: 2, Milwaukee, WI, Summer 1999.

F. W. Breyfogle III, "Deploy Six Sigma Plan as a Four-Phase Project," ASQ Quality Management Forum, 25, 3, Milwaukee, WI, Fall 1999.

F. W. Breyfogle III, J. M. Cupello, B. Meadows, Managing Six Sigma: A Practical Guide to Understanding, Assessing, and Implementing the Strategy that Yields Bottom-line Success, Wiley, New York, NY, 2000.

P. M. Senge, The Fifth Discipline: The Art and Practice of the Learning Organization, Doubleday/Current, New York, 1990.

M. J. Harry, "The Nature of Six Sigma Quality," Technical Report, Government Electronics Group, Motorola, Inc., Scottsdale, AZ.

Six Sigma, Basic Steps & Implementation

Breyfogle III, Forrest W. (1999) *Implementing Six Sigma: Smarter Solutions using Statistical Methods*, Wiley, New York, NY.

Jones D. (1998), Firms Air for Six Sigma Efficiency, *USA Today*, 7/21/98 Money Section Lowe, J. (1998), *Jack Welch Speaks*, Wiley, New York.

Spagon, P (1998), personal communications.

Wiggenhorn, B (1999), Except from foreword in *Implementing Six Sigma: Smarter Solutions using Statistical Methods*, Forrest W. Breyfogle III Wiley, New York, NY 1999.

Distribution	Functions	Density/Probability/Function
Beta	(x,a,b) (x,a,b) (p,a,b) (a,b)	$$f(x,a,b) = \frac{x^{a-1}(1-x)^{b-1}}{B(a,b)}$$ for $0 \le x \le 1$ and $a,b>0$.
Binomial	(x,n,p) (x,n,p) (s,n,p) (n,p)	$$\Pr(x,n,p) = \binom{n}{x} p^x (1-p)^{n-x}$$ if $x = 0,1,\dots,n$, and 0 otherwise, for $0 \le p \le 1$
Chi-square	(x,v) (x,v) (p,v) (v)	$$f(x,v) = \frac{1}{2^{v/2}\,\Gamma(v/2)}\, x^{v/2-1} e^{-x/2}$$ where $x \ge 0$, and $v>0$. Note that the degrees-of-freedom parameter v need not be an integer.
Exponential	(x,m) (x,m) (p,m) (m)	$$f(x,m) = \frac{1}{m} e^{-x/m}$$ for $x \ge 0$, and $m>0$.
Extreme Value (Type I - minimum)	(x) (x) $(p),$	$$f(x) = \exp(x - e^x)$$ for $-\infty < x < \infty$.

	(p)	
F-distribution	(x,v1,v2) (x,v1,v2) (p,v1,v2) (v1,v2)	$$f(x, v_1, v_2) = \frac{v_1^{v_1/2} v_2^{v_2/2}}{B(v_1/2, v_2/2)} x^{(v_1-2)/2}$$ $$\cdot (v_2 + v_1 x)^{-(v_1+v_2)/2}$$ where $x \geq 0$, and $v_1, v_2 > 0$. *Note that the functions allow for fractional degrees-of-freedom parameters.*
Gamma	(x,r,s) (x,r,s) (p,r,s) (r,s)	$$f(x, b, r) = b^{-r} x^{r-1} e^{-x/b} / \Gamma(r)$$ where $x \geq 0$, $r>0$, and $r>0$.
Laplace	(x) (x) (x)	$$f(x) = \frac{1}{2} e^{-|x|}$$ for $-\infty < x < \infty$.
Logistic	(x) (x) (p)	$$f(x) = \frac{1}{1 + e^{-x}} = \frac{e^x}{1 + e^x}$$ for $-\infty < x < \infty$.
Log-normal	(x,m,s) (x,m,s) (p,m,s) (m,s)	$$f(x, m, s) = \frac{1}{x\sqrt{2\pi s^2}} e^{-(\log x - m)^2 / 2s^2}$$ $x>0$, $-\infty < m < \infty$, and $s>0$. *Note that exp(m) is the median of the lognormal.*
Negative Binomial	(x,n,p) (x,n,p) (s,n,p) (n,p)	$$\Pr(x, n, p) = \frac{\Gamma(x + n)}{\Gamma(x + 1)\Gamma(n)} p^n (1 - p)^x$$ if $x = 0, 1, \ldots, n, \ldots$, and 0 otherwise, for $0 \leq p \leq 1$

Normal (Gaussian)	(x)	$f(x) = (2\pi)^{-1/2} \varepsilon^{-x^2/2}$
	(x)	for $-\infty < x < \infty$.
	(p)	

Pareto	(x,a,k)	$f(x, a, k) = \dfrac{ak^a}{x^{a+1}}$
	(x,a,k)	
	(p,a,k)	for a>0, and $x \geq k \geq 0$.
	(a,k)	

Poisson	(x,m)	$\Pr(x, m) = m^x e^{-m}/x!$
	(x,m)	if $x = 0, 1, \ldots, n, \ldots$, and 0 otherwise, for
	(p,m)	m>0.
	(m)	

t-distribution	(x,v)	$f(x, v) = \dfrac{\Gamma((v + 1)/2)}{(v\pi)^{1/2}\,\Gamma(v/2)}\left(1 + (x^2/v)\right)^{-(v+1)}$
	(x,v)	
	(p,v)	for $-\infty < x < \infty$, and v>0.
	(v)	Note that v=1 is the Cauchy distribution.

Uniform	(x,a,b)	$f(x) = \dfrac{1}{b - a}$
	(x,a,b)	
	(p,a,b)	for $a \leq x \leq b$.
	(a,b), rnd	

Weibull	(x,m,a)	$f(x, m, a) = a m^{-a} x^{a-1} e^{-(x/m)^a}$
	(x,m,a)	where x>0 and m,a>0.
	(p,m,a)	
	(m,a)	

Application of statistical tools for DMAIC phases

Six Sigma Tool Name	Define	Measure	Analyze	Improve	Control	Plan and Define	Product Design	Process Design	Validation	Feedback & Corrective
Affinity Diagram	X					X				
Brainstorming			X	X		X	X			
Business Case	X					X				
Cause and Effect Diagrams			X						X	X
Charter	X						X	X		
Consensus			X					X	X	X
Control Charts		X	X	X	X		X	X	X	X
Critical to Quality (CTQ)	X					X	X	X	X	X
Data Collection		X	X	X	X				X	X
Design of Experiments			X	X			X	X		
Flow Diagrams	X	X	X	X	X		X	X		
Frequency Plots			X	X	X		X			
FMEA		X		X			X			
Gage R & R		X					X	X	X	X
Hypothesis Testing			X				X	X	X	
Kano Diagram		X				X	X	X	X	X
Planning Tools - Gantt Chart				X		X	X	X	X	X
Pareto Charts		X	X	X		X				
Prioritization Matrix		X		X			X	X	X	X
Process Capability		X		X				X	X	
Process Sigma		X		X			X	X	X	X
Quality Control Process Charts					X			X	X	X
Regression Analysis			X			X	X			
Rolled Throughput Yield	X						X	X		
Sampling		X	X	X	X	X	X			

	C1	C2	C3	C4	C5	C6	C7	C8	C9	C10
Scatter Plots			X				X	X	X	X
SIPOC (High-level flowchart)	X								X	
Stakeholder Analysis	X			X			X	X		
Standardization					X	X	X	X		X
Time Series Plots (Run Charts)		X								
VOC (Voice of the Customer)	X									

FMEA

Worksheet

Process Step	Key Process Input	Failure Modes - What can go wrong?	Effects	Causes	Current Controls

Six Sigma Reference Table

DPM	Sigma Short Term	Sigma Long Term	Yield	Cpk
2	6	4.5	99.99966	2
5	5.9	4.4	99.99954	1.97
9	5.8	4.3	99.99915	1.93
13	5.7	4.2	99.9987	1.9
21	5.6	4.1	99.9979	1.87
32	5.5	4	99.9968	1.83
48	5.4	3.9	99.995	1.8
72	5.4	3.9	99.993	1.77
108	5.2	3.7	99.989	1.73
159	5.1	3.6	99.984	1.7
233	5	3.5	99.98	1.67
337	4.9	3.4	99.97	1.63
483	4.8	3.3	99.95	1.6
687	4.7	3.2	99.93	1.57
968	4.6	3.1	99.90	1.53
1,350	4.5	3	99.87	1.5
1,866	4.4	2.9	99.81	1.47
2,555	4.3	2.8	99.74	1.43
3,467	4.2	2.7	99.65	1.4
4.661	4.1	2.6	99.5	1.37
6,210	4	2.5	99.4	1.33

8,198	3.9	2.4	99.2	1.3
10,724	3.8	2.3	98.9	1.27
13,903	3.7	2.2	98.6	1.23
17,864	3.6	2.1	98.2	1.2
22,750	3.5	2	97.7	1.17
28,716	3.4	1.9	97.1	1.13
35,930	3.3	1.8	96.4	1.1
44,565	3.2	1.7	95.5	1.07
54,799	3.1	1.6	94.5	1.03
66,807	3	1.5	93.3	1
80,757	2.9	1.4	91.9	0.97
96,801	2.8	1.3	90.3	0.93
115,070	2.7	1.2	88.5	0.9
135,666	2.6	1.1	86.4	0.87
158,655	2.5	1	84.1	0.83
184,060	2.4	0.9	81.6	0.8
211,855	2.3	0.8	78.8	0.77
241,964	2.2	0.7	75.8	0.73
274,253	2.1	0.6	72.6	0.7
308,538	2	0.5	69.1	0.67
344,578	1.9	0.4	65.5	0.63
382,089	1.8	0.3	61.8	0.6
420,740	1.7	0.2	57.9	0.57
460,172	1.6	0.1	54.0	0.53
500,000	1.5	0	50.0	0.5

539,828	1.4	-0.1	46.0	0.47
579,260	1.3	-0.2	42.1	0.43
617,911	1.2	-0.3	38.2	0.4
655,422	1.1	-0.4	34.5	0.37
691,462	1	-0.5	30.9	0.33
725,747	0.9	-0.6	27.4	0.30
758,036	0.8	-0.7	24.2	0.27
788,145	0.7	-0.8	21.2	0.23
815,940	0.6	-0.9	18.4	0.20
841,345	0.5	-1	15.9	0.17
864,334	0.4	-1.1	13.6	0.13
884,930	0.3	-1.2	11.5	0.10
903,199	0.2	-1.3	9.7	0.07
919,243	0.1	-1.4	8.1	0.03
933,193	0	-1.5	6.7	0.00

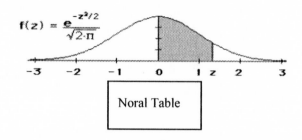

$$f(z) = \frac{e^{-z^2/2}}{\sqrt{2\cdot\pi}}$$

Noral Table

z	0.00	0.01	0.02	0.03	0.04	0.05	0.06	0.07	0.08	0.09
0.0	0.0000	0.0040	0.0080	0.0120	0.0160	0.0199	0.0239	0.0279	0.0319	0.0359
0.1	0.0398	0.0438	0.0478	0.0517	0.0557	0.0596	0.0636	0.0675	0.0714	0.0753
0.2	0.0793	0.0832	0.0871	0.0910	0.0948	0.0987	0.1026	0.1064	0.1103	0.1141
0.3	0.1179	0.1217	0.1255	0.1293	0.1331	0.1368	0.1406	0.1443	0.1480	0.1517
0.4	0.1554	0.1591	0.1628	0.1664	0.1700	0.1736	0.1772	0.1808	0.1844	0.1879
0.5	0.1915	0.1950	0.1985	0.2019	0.2054	0.2088	0.2123	0.2157	0.2190	0.2224
0.6	0.2257	0.2291	0.2324	0.2357	0.2389	0.2422	0.2454	0.2486	0.2517	0.2549
0.7	0.2580	0.2611	0.2642	0.2673	0.2704	0.2734	0.2764	0.2794	0.2823	0.2852
0.8	0.2881	0.2910	0.2939	0.2967	0.2995	0.3023	0.3051	0.3078	0.3106	0.3133
0.9	0.3159	0.3186	0.3212	0.3238	0.3264	0.3289	0.3315	0.3340	0.3365	0.3389
1.0	0.3413	0.3438	0.3461	0.3485	0.3508	0.3531	0.3554	0.3577	0.3599	0.3621
1.1	0.3643	0.3665	0.3686	0.3708	0.3729	0.3749	0.3770	0.3790	0.3810	0.3830
1.2	0.3849	0.3869	0.3888	0.3907	0.3925	0.3944	0.3962	0.3980	0.3997	0.4015
1.3	0.4032	0.4049	0.4066	0.4082	0.4099	0.4115	0.4131	0.4147	0.4162	0.4177
1.4	0.4192	0.4207	0.4222	0.4236	0.4251	0.4265	0.4279	0.4292	0.4306	0.4319
1.5	0.4332	0.4345	0.4357	0.4370	0.4382	0.4394	0.4406	0.4418	0.4429	0.4441

1.6	0.4452	0.4463	0.4474	0.4484	0.4495	0.4505	0.4515	0.4525	0.4535	0.4545
1.7	0.4554	0.4564	0.4573	0.4582	0.4591	0.4599	0.4608	0.4616	0.4625	0.4633
1.8	0.4641	0.4648	0.4656	0.4664	0.4671	0.4678	0.4685	0.4692	0.4699	0.4706
1.9	0.4713	0.4719	0.4725	0.4732	0.4738	0.4744	0.4750	0.4756	0.4761	0.4767
2.0	0.4772	0.4778	0.4783	0.4788	0.4793	0.4798	0.4803	0.4808	0.4812	0.4817
2.1	0.4821	0.4826	0.4830	0.4834	0.4838	0.4842	0.4846	0.4850	0.4854	0.4857
2.2	0.4861	0.4864	0.4868	0.4871	0.4875	0.4878	0.4881	0.4884	0.4887	0.4890
2.3	0.4893	0.4896	0.4898	0.4901	0.4904	0.4906	0.4909	0.4911	0.4913	0.4916
2.4	0.4918	0.4920	0.4922	0.4925	0.4927	0.4929	0.4931	0.4932	0.4934	0.4936
2.5	0.4938	0.4940	0.4941	0.4943	0.4945	0.4946	0.4948	0.4949	0.4951	0.4952
2.6	0.4953	0.4955	0.4956	0.4957	0.4959	0.4960	0.4961	0.4962	0.4963	0.4964
2.7	0.4965	0.4966	0.4967	0.4968	0.4969	0.4970	0.4971	0.4972	0.4973	0.4974
2.8	0.4974	0.4975	0.4976	0.4977	0.4977	0.4978	0.4979	0.4979	0.4980	0.4981
2.9	0.4981	0.4982	0.4982	0.4983	0.4984	0.4984	0.4985	0.4985	0.4986	0.4986
3.0	0.4987	0.4987	0.4987	0.4988	0.4988	0.4989	0.4989	0.4989	0.4990	0.4990

t table with right tail probabilities

df\p	0.40	0.25	0.10	0.05	0.025	0.01	0.005	0.0005
1	0.324920	1.000000	3.077684	6.313752	12.70620	31.82052	63.65674	636.6192
2	0.288675	0.816497	1.885618	2.919986	4.30265	6.96456	9.92484	31.5991
3	0.276671	0.764892	1.637744	2.353363	3.18245	4.54070	5.84091	12.9240
4	0.270722	0.740697	1.533206	2.131847	2.77645	3.74695	4.60409	8.6103
5	0.267181	0.726687	1.475884	2.015048	2.57058	3.36493	4.03214	6.8688
6	0.264835	0.717558	1.439756	1.943180	2.44691	3.14267	3.70743	5.9588
7	0.263167	0.711142	1.414924	1.894579	2.36462	2.99795	3.49948	5.4079
8	0.261921	0.706387	1.396815	1.859548	2.30600	2.89646	3.35539	5.0413
9	0.260955	0.702722	1.383029	1.833113	2.26216	2.82144	3.24984	4.7809
10	0.260185	0.699812	1.372184	1.812461	2.22814	2.76377	3.16927	4.5869
11	0.259556	0.697445	1.363430	1.795885	2.20099	2.71808	3.10581	4.4370
12	0.259033	0.695483	1.356217	1.782288	2.17881	2.68100	3.05454	4.3178
13	0.258591	0.693829	1.350171	1.770933	2.16037	2.65031	3.01228	4.2208
14	0.258213	0.692417	1.345030	1.761310	2.14479	2.62449	2.97684	4.1405
15	0.257885	0.691197	1.340606	1.753050	2.13145	2.60248	2.94671	4.0728
16	0.257599	0.690132	1.336757	1.745884	2.11991	2.58349	2.92078	4.0150
17	0.257347	0.689195	1.333379	1.739607	2.10982	2.56693	2.89823	3.9651
18	0.257123	0.688364	1.330391	1.734064	2.10092	2.55238	2.87844	3.9216
19	0.256923	0.687621	1.327728	1.729133	2.09302	2.53948	2.86093	3.8834
20	0.256743	0.686954	1.325341	1.724718	2.08596	2.52798	2.84534	3.8495
21	0.256580	0.686352	1.323188	1.720743	2.07961	2.51765	2.83136	3.8193
22	0.256432	0.685805	1.321237	1.717144	2.07387	2.50832	2.81876	3.7921
23	0.256297	0.685306	1.319460	1.713872	2.06866	2.49987	2.80734	3.7676
24	0.256173	0.684850	1.317836	1.710882	2.06390	2.49216	2.79694	3.7454
25	0.256060	0.684430	1.316345	1.708141	2.05954	2.48511	2.78744	3.7251

26	0.255955	0.684043	1.314972	1.705618	2.05553	2.47863	2.77871	3.7066
27	0.255858	0.683685	1.313703	1.703288	2.05183	2.47266	2.77068	3.6896
28	0.255768	0.683353	1.312527	1.701131	2.04841	2.46714	2.76326	3.6739
29	0.255684	0.683044	1.311434	1.699127	2.04523	2.46202	2.75639	3.6594
30	0.255605	0.682756	1.310415	1.697261	2.04227	2.45726	2.75000	3.6460
inf	0.253347	0.674490	1.281552	1.644854	1.95996	2.32635	2.57583	3.2905

Right tail areas for the *Chi-square* Distribution

df\area	.995	.990	.975	.950	.900	.750	.500	.250	.100	.050	.025	.010	.005
1	0.00004	0.0001	0.0009	0.0039	0.0157	0.1015	0.4549	1.3233	2.7055	3.8414	5.0238	6.6349	7.8794
2	0.01003	0.02010	0.05064	0.10259	0.21072	0.57536	1.38629	2.77259	4.60517	5.99146	7.37776	9.21034	10.59663
3	0.07172	0.11483	0.21580	0.35185	0.58437	1.21253	2.36590	4.10834	6.25139	7.81473	9.34840	11.34487	12.83816
4	0.20699	0.29711	0.48442	0.71072	1.06362	1.92256	3.35669	5.38527	7.77944	9.48773	11.14329	13.27670	14.86026
5	0.41174	0.55430	0.83121	1.14548	1.61031	2.67460	4.35146	6.62568	9.23636	11.07050	12.83250	15.08627	16.74960
6	0.67573	0.87209	1.23734	1.63538	2.20413	3.45460	5.34812	7.84080	10.64464	12.59159	14.44938	16.81190	18.54758
7	0.98926	1.23904	1.68987	2.16735	2.83311	4.25485	6.34581	9.03715	12.01704	14.06714	16.01276	18.47531	20.27774
8	1.34441	1.64650	2.17973	2.73264	3.48954	5.07064	7.34412	10.21885	13.36157	15.50731	17.53455	20.09024	21.95495
9	1.73493	2.08790	2.70039	3.32511	4.16816	5.89883	8.34283	11.38875	14.68366	16.91898	19.02277	21.66599	23.58935
10	2.15586	2.55821	3.24697	3.94030	4.86518	6.73720	9.34182	12.54886	15.98718	18.30704	20.48318	23.20925	25.18818
11	2.60322	3.05348	3.81575	4.57481	5.57778	7.58414	10.34100	13.70069	17.27501	19.67514	21.92005	24.72497	26.75685
12	3.07382	3.57057	4.40379	5.22603	6.30380	8.43842	11.34032	14.84540	18.54935	21.02607	23.33666	26.21697	28.29952
13	3.56503	4.10692	5.00875	5.89186	7.04150	9.29907	12.33976	15.98391	19.81193	22.36203	24.73560	27.68825	29.81947
14	4.07467	4.66043	5.62873	6.57063	7.78953	10.16531	13.33927	17.11693	21.06414	23.68479	26.11895	29.14124	31.31935
15	4.60092	5.22935	6.26214	7.26094	8.54676	11.03654	14.33886	18.24509	22.30713	24.99579	27.48839	30.57791	32.80132
16	5.14221	5.81221	6.90766	7.96165	9.31224	11.91222	15.33850	19.36886	23.54183	26.29623	28.84535	31.99993	34.26719
17	5.69722	6.40776	7.56419	8.67176	10.08519	12.79193	16.33818	20.48868	24.76904	27.58711	30.19101	33.40866	35.71847
18	6.26480	7.01491	8.23075	9.39046	10.86494	13.67529	17.33790	21.60489	25.98942	28.86930	31.52638	34.80531	37.15645
19	6.84397	7.63273	8.90652	10.11701	11.65091	14.56200	18.33765	22.71781	27.20357	30.14353	32.85233	36.19087	38.58226
20	7.43384	8.26040	9.59078	10.85081	12.44261	15.45177	19.33743	23.82769	28.41198	31.41043	34.16961	37.56623	39.99685
21	8.03365	8.89720	10.28290	11.59131	13.23960	16.34438	20.33723	24.93478	29.61509	32.67057	35.47888	38.93217	41.40106
22	8.64272	9.54249	10.98232	12.33801	14.04149	17.23962	21.33704	26.03927	30.81328	33.92444	36.78071	40.28936	42.79565
23	9.26042	10.19572	11.68855	13.09051	14.84796	18.13730	22.33688	27.14134	32.00690	35.17246	38.07563	41.63840	44.18128
24	9.88623	10.85636	12.40115	13.84843	15.65868	19.03725	23.33673	28.24115	33.19624	36.41503	39.36408	42.97982	45.55851
25	10.51965	11.52398	13.11972	14.61141	16.47341	19.93934	24.33659	29.33885	34.38159	37.65248	40.64647	44.31410	46.92789
26	11.16024	12.19815	13.84390	15.37916	17.29188	20.84343	25.33646	30.43457	35.56317	38.88514	41.92317	45.64168	48.28988
27	11.80759	12.87850	14.57338	16.15140	18.11390	21.74940	26.33634	31.52841	36.74122	40.11327	43.19451	46.96294	49.64492
28	12.46134	13.56471	15.30786	16.92788	18.93924	22.65716	27.33623	32.62049	37.91592	41.33714	44.46079	48.27824	50.99338
29	13.12115	14.25645	16.04707	17.70837	19.76774	23.56659	28.33613	33.71091	39.08747	42.55697	45.72229	49.58788	52.33562
30	13.78672	14.95346	16.79077	18.49266	20.59923	24.47761	29.33603	34.79974	40.25602	43.77297	46.97924	50.89218	53.67196

About the Author

Fred Soleimannejed has obtain his master degree in quality assurance from San Jose State University. He has worked for fortune 500 companies since 1988. He has held executive positions and global quality manager at different companies. He has worked on the Six Sigma Concept since 1995.

Printed in the United States
18616LVS00005B/32